Drought in India
Assessment and Impact

S. B Golahit
Reader & Head
Department of Economics
S.P.D.M College, Shirpur, Maharashtra
(Dist.: Dhule)

AKHAND

AKHAND PUBLISHING HOUSE
NEW DELHI (INDIA)

Published by
AKHAND PUBLISHING HOUSE
L-9 A, IST FLOOR, STREET NO. 42
SADATPUR EXTENSION
DELHI-110094 (INDIA)
CELL- 0-9013387535, 0-9968628081

Drought in India: Assessment and Impact

Edition 2010
ISBN 978-81-910525-1-0

Printed in India
at Asian Offset Printers, Delhi

Preface

Drought has been constant visitor to India due to its tropical situation and erratic and uneven behaviour of Indian monsoon.

Drought and Floods are common features of the economy of the country. Before independence the country is facing the worst drought of the period: The widespread drought has upset the country's programme of self-sufficiency in food grains. The drought had been most severe and persists in many states. The Central Government is trying to provide necessary financial assistance to drought affected areas. All possible remedial measures are being adopted by Government.

The book is divided into seven chapter. The analysis, information and statistics provide a new insight to the students of economics as well as lay readers. My debt to those who have helped me in one way or other is heavy indeed. While I take this opportunity to thank all of them—they are too numerous to be mentioned in a brief preface, I would like to acknowledge my deep sense of gratitude to Prof. R. P. Rajput for his help at all stages.

I am grateful to all those who helped me in the completion of this book.

While I owe a debt, of gratitude to a large number of friends and well-wishers, the opinions expressed in this book are entirely mine and I alone take the responsibility for shortcomings, if any. I also express my sincere thanks to my Adhyayan Publishers & Distributors and their efficient editorial staff who helped in publishing this book in time.

Lastly, I will be failing in my duty if I do not mention here the tremendous co-operation I received from my wife Mayadevi in the completion of this voluminous work, in particular, whose patience, encouragement, understanding and have helped in bringing this effort to fruition.

Dr. S.B. Golahit

April 2010

Contents

Introduction

Drought has been constant visitor to India due to its tropical situation and erratic and uneven behaviour of Indian monsoon. Crops have been subject to failure either partly or completely resulting in famine. That means, famine is not a new feature in the history of India, In the country, famines had occurred in the past centuries, and tradition and history have recorded gruesome tales of the trail of misery which famines had left behind them. During the British rule, India has had 22 famines excluding severe scarcity. The immediate cause of famine in almost every instance is the failure of rains and this cause will continue to operate until we have a more extensive system of irrigation. The problem of drought is so acute that every year, some parts of the country suffer from drought. Some time drought is so extreme in its intensity and spread that it leads to a complete failure of crops. Agriculture is the main source of livelihood in the country. It contributes about 30 per cent of the national income, supplies raw materials for a number of ,important industries and provides products for about 20 per cent of the country's foreign exchange earnings. The total area of India is 330 million hectares of this 142 million consists of sown area. Of the total sown area only about 30 per cent is under irrigation. The disturbing fact is that in spite of 40 years of planned development of irrigation about 70 per cent of the sown area is still dependent on rainfall. Agriculture provides livelihood to about 70 per cent of the country's population, peoples dependence on agriculture is such that when agriculture fails, the economy fails and the people particularly the poor starve. Common people suffer due to scarcity of water, fodder, cereals, pulses and edible oils and due to the rise in prices. People have to move from one place to another in search of employment. An adverse impact of drought or famine on agriculture produces a chain reaction on all the other sectors of the economy. Food production declines, employment declines, livestock succumbs.

It is contended that sources of livelihood for the rural population are getting diversified, and in some areas the rapid advance of productivity in agriculture has substantially raised the general living standards and reduced the insecurity of rural life; in many states a wide array of income support schemes provide a measure of protection against destitution to poor households, and by some accounts at least a discernible trend towards decreasing poverty has emerged since the mid sixties.

But even these recent changes in economic opportunities and policies do not seem to be adequate for prevention of famines. In the non-irrigated, semi-arid parts of India stagnation or near stagnation of yields, population pressure and the increasing frequency of droughts keep the rural population at the mercy of the monsoon, the vulnerability of impoverished classes remains and the need persists for a very extensive and expensive relief system.

In this dissertation, literature on economic causes and impact of famine in India is reviewed. In first chapter, causes of famines under the British rule are examined. In the second chapter economic impact of famine or droughts is discussed briefly. In the third chapter, efforts for drought mitigation in the post-1947 period are reviewed. Famine relief measures which include new approach for relief financing are discussed in the fourth chapter. In chapter fifth assessment and impact of droughts 1987 and 2002, are dicussed. In the sixth chapter drought relief measures have been discussed briefly. In the last chapter, conclusions are presented.

1

Famines in India during British Rule

Numerous famines occurred in India throughout the nineteenth century, and their victims were often counted in millions. There is a fair amount of agreement among nineteenth century analysts and later economics historians concerning the proximate causes underlying these catastrophes. In most cases, famine followed massive crop failures resulting from drought. The immediate effect of these crop failure was not only to reduce food availability in the affected region, but also and more importantly to disrupt rural economy.

There are many causes of famines i.e., natural, political, social and economic. But in this book only the economic causes of Indian Famines are considered. This Chapter is divided into two parts. Part-I outlines the economic condition of India in the 19th century. Part-II describes the economic causes of famine under the British rule.

Following number of famines were before British rule in India.

In Century	*No. of famines*
11th	2
13th	1
14th	3

In Century	*No. of famines*
15th	2
16th	3
17th	3
18th	4

Source: Williom Digby, Prosperous, British India London: T. Fisher Unwin, 1901, p. 123.

Economic Condition of India in the 19th century

India is a land rich in resources in comparison to most other lands in the world.[1] In fact, it is a rich country inhabited by the poor. But the conditions were different in ancient times, when it was famous for prosperity. Even as late as the 17th century, it was the "agricultural mother of Asia and the industrial workshop of the world".[2] The process of economic deterioration started in the 18th century and it reached its culmination under the rule of the East India Company. The strain of the East India Company's policy of exploitation of the Indian provinces and the decay of the cottage industries shattered the Indian economy. It was further weakened due to competition from British machine made goods.[3]

The East India Company's monopoly over Indian trade, exemption of British merchants from the payment of the usual customs and transit duties, cheap imitation of Indian goods by British manufactures, heavy duties imposed on Indian goods exported to England, payment of fat salaries to British officials, military expenditure on imperial expansion, annual drain of £1,50,00,000 for home charges and various other methods adopted by the British Government, crippled the Indian economy by the middle of the 19th century. Not only textiles but also other indigenous industries like jewelry, metal work, ivory carving and paper making were ruined.[4]

All those factors contributed to the progressive decline of the Indian economy. The standard of living in the early 18th century witnessed so much deterioration that we find an appreciable fall in the consumption of clothes, blankets, butter oil, salt and sugar and in the general level of living of the common man.[5]

Lord Lawrence writing in 1864 as Viceroy of India admitted the above fact. He observed "India is on the whole a very poor country; the mass of population enjoys only a scanty subsistence. No wonder that the Indian people suffered slightest failure of rain and died in thousands."[6]

Indian economy in the early 19th century had the potential for moving on to the modern stage of economic development. Unfortunately this did not happen. On the contrary, it moved backwards and got stuck in a state of economic stagnation. This was the period during which Western European countries—the USA, Soviet Union, Japan, etc. made rapid strides with their national governments very active. India, on the other hand, was pushed back largely because of foreign rule. Despite similar initial conditions, the distance between advancing countries and India lengthened. In course of time, Indian economy acquired the traits of backward one.

The backward movement of the Indian economy is symbolized by the down-ward sliding of the per capita income. Among the many estimates, I refer to those made by Dadabhai Naoroji, the pioneer in the field, William Digby, Atkinson, Baring and Harbour and Lord Curzon, with all the limitations of data collection by individuals, these are the best available estimates. These estimates may not be strictly comparable on account of differences in methodology and personal biases of the estimators, yet these do indicate broad trends when put together and suitably adjusted for the purpose. Table presents these estimates for the second half of 19th century.

For purposes of comparison over time, the data has been split into three periods, each consisting or proximate years; the first period is 1867-68 and 1875, the second 1880 and 1882, and the third 1875, 1899 and 1901. The per capitas income of each period is the average of the estimates of the years falling in the period. On this basis the other figures have been calculated.

Table 1.2
Per Capita Income [1867-68–1901]

Years	Estimated by		Price capita Income [Rs]	Price of paddy (average of Madras and PU price) Rs per maund	Per Capita income in terms of paddly maund	Per capita income (at 1873 price)
1867-68	Naoroji	20.00	25.00	1-10-0	15.4	24.35
1875	Atkinson	30.5				
1880	Digby	22.5	25.00	1-10-0	15.4	24.05
1882	Baring & Barbour	27.0				
1895	Atkinson	39.5	29.00	2-6-0	12.7	23.13
1899	Digby	18.0				
1901	Curzon	30.0				

Source: V.V. Bhatt, Aspects of Economic Change and Policy in India, p. 19.

From this it is clear that the per capita income in terms of paddy was around 15 maunds in the first period which remained constant in the second period followed by a decline of 17 per cent during the third period. In money terms the per capita income at constant prices shows a decline for all these periods, moving down from Rs. 24.35 to 24.05 and then to Rs. 23.13.

For the period 1900-1947 the per capita income estimates by Surendra J. Patel and K. Mukerji are given in Table 1.3.

Table 1.3
Changes in Per Capita Income [1905-6–1945-46]
[1905-1906–1915-16=100]

Years	*Surendra Patel*	*Mukerji*
1905-06 to 1915-16	100	100
1916-17 to 1925-26	103.9	110
1926-27 to 1935-36	98.3	112
1936-37 to 1945-46	91.1	112

Source: V.V. Bhatt, *op. cit,* p. 20.

According to Patel's estimate the real per capita income moved down by about nine per cent during 1905-06 to 1945-46. Mukerjii's estimates, on the other hand, show an upward trend–a rise of 12 per cent during this period. But this rise is largely statistical because there is an element of overestimation in it. In this estimate two questionable assumptions have been made. The rate of growth of income from professions, domestic services and house property has been assumed to be equal to that of income from industrial production. Further, with regard to the items under "other commerce" and "small enterprises" it was assumed that the rate of increase in incomes was the same as that for commercial activity. These assumptions tended to inflate their contributions. If we allow for these, the actual rise in per capita income would certainly be much les, if not negative.

From the above mentioned estimates, it is obvious that

during the 100 years of British rule per capita income remained stagnant, with a tilt towards the downward side. In such circumstances, it was natural that famines were more frequent and severe under the British rule. Dadabhai Naoroji aptly described the British government as one of the most 'extortionate and oppressive' that ever existed in India."[7]

An English writer has discussed the problems and wrote that the Government of India was and always has been an agency of "commercial exploitation maintained by military force".[8]

The Economic Causes of Indian Famines under the British Rule

During the British rule (1858-1947), India has had 22 famines excluding severe scarcities.[9] Primary causes of famines was failure of rains in a large part of the country. The crops were, therefore, badly damaged causing failure in kharif crop yields ranging between 45 and 80 per cent in different provinces. The famine of 1896-97 was widespread, and was officially described as the most disastrous famine of the country. In the Central Provinces the crop yield for the year was estimated at 45 per cent of the normal; in North Western provinces and Oudh 60 per cent; in Bengal 67 per cent; in Bombay 65 per cent; in the Punjab 75 per cent and in Madras 80 per cent.[10]

Overall loss of crop was estimated at 33 per cent of the average annual production. Heavy export of grains during the previous (1895) period in Bombay, Bengal and Madras drained away all the surpluses and thus the situation was aggravated in these provinces. The country was confronted with food shortage of a great magnitude, unique in the history of its famines. The first intimation of famine came to government from North-Western provinces at the end of September 1896.[11] Famine took a violent turn next month. There were grain riots in Delhi, Agra, Nagpur, Muzaffarnagar and Mhow.[12]

Commercialisation of Agriculture

This is one of the causes of Indian famines. The establishment of the British rule brought about certain fundamental changes in the economic structure. Till the end of the first half of the 18th century the Indian villages were nearly self-sufficient. But commercialization introduced by the British resulted in the large scale substitution of commercial crops like indigo in Bihar and Bengal and cotton in the north and west, replacing food crops like jowar, bajra and pulses. This led to creation of local scarcities which owing to transport difficulties often assumed the dimensions of famine.[13]

Decline of Cottage Industries

An impression persists that India has "always been an agriculture country and that the people of India are by nature and tradition deficient in 'industrial and commercial enterprises..."[14]

The fact of the matter is that the flooding of cheap machine made imports from abroad aided by railway development not only undermined the competitive power of the urban industries like cotton and silk textiles, woollen cloth, brass and copper wares, but also destroyed the livelihood of the village potter, the oil expeller, and the shoe maker. Craftsman and artisans both in urban and rural areas were thus thrown out of their traditional occupations and had to turn to agriculture. In this respect the contrast between India and the countries of the west is striking. In England and other countries in the 19th century rural population was released from agriculture to provide wage labour to the expanding industries. In India manufacturing sector threw out labour to be absorbed in stagnant agriculture. This labour was exposed to the uncertainties of employment due not .only to the seasonal character of Indian agriculture, but also because of the vagaries of rainfall. Large population therefore, became highly vulnerable to famine.

Land Revenue Policy

Land revenue policy of British Government was oppressive. The burden of revenue was not equitable. There were wide variations from province to province and even within the same province. For example, land revenue per head of population ranged from about 0.56 paise in Bengal to Rs. 2/- in Bombay, Rs, 2.56 in Sind, Rs. 1.60 in Madras and Rs, 1.30 in Punjab.[15] Louis Mallet, taking the figures for 1871 calculated the net land revenue per square mile and per head of population in the different provinces.[16] (Table 1.4).

Table 1.4
Inequality in Land Revenue

Province	*Land revenue per sq. mile*			*Land revenue per head of population*		
	£	s	d	£	s	d
Bengal & Assam	17	14	3	0	1	1½
Bengal & Sind	22	0	11	0	3	4½
Madras	29	0	9	0	2	6¾
H.W.P.	45	4	9	0	2	4¾
Punjab	17	6	4	0	2	0

Source: Famine Commission Report, 1880, Appendix I, p. 143.

Baird Smith pointed out the relation between land revenue settlement and famines, and emphasized that an oppressive revenue demand was the direct cause of famines in India.[17] R. C. Dutta maintained that though Indian famines were caused by the failure of rain, their frequency and intensity was due to the chronic poverty of the cultivator caused by over assessment of the land revenue.[18]

In the permanently settled and zamindari areas, the landlords oppressed the peasants in several ways and the government had to pass several laws for the protection of the tenants from rack-renting and eviction. In temporarily settled areas there were revisions of settlements,[19] and the revenue demand was often increased. For example, according to R. C. Dutta, gross demand increased in Madras, between 1876 and 1898, by over 70 per cent though the increase in the area of cultivation was hardly 14 per cent.[20] Similar increase was noticeable in other parts as well.[21] Thus, Baird Smith mentions that, "the assessments were excessive, unequal and unintelligent".[22]

Thus the land revenue relations which embodied the creation of a class of land owners and a class of cultivators, separated ownership from cultivation. The landlords were interested in extracting high rents, leaving a pittance with the cultivators. The investment on land fell sharply because the cultivators had to part with a major portion of the produce in the form of the rent of land to the landlords and interest to the moneylenders. The land revenue policy of the government created serious difficulties for the cultivators and was one of the basic causes of suffering in times of famines."[23]

Revenue collection in British period differed from that in the earlier periods in two important ways. It was collected regularly and vigorously, as administrative efficiency increased, and it was collected in cash and was related to land and not to the level of production. In the earlier period of British rule, in many cases, if the revenue was not paid before sunset on the last date of payment the estate was sold. The zamindars thus turned upon the peasants who had in their turn no option but to borrow money from the money-lender who gave it on his own terms. The peasants were illiterate and the revenue collectors are reported to have collected more than the actual demand which they pocketed."[24]

The provincial report of the N. W. Famine of 1876-79 mentions that wherever the rents were *batai*, the people, as a rule, were less afflicted than those where rents were paid in cash. Therefore the large number of sales of land and mortgages in 1896-98 and 1899-1900 reveal that the peasant had often no option but to sell or mortgage his land to pay the revenue.

Growing Indebtedness

The Deccan Riots Commission found in 1875 that 1/3 of the occupants of government land were' in debt, that the average debt per occupant was Rs. 371 and that this debt covered 18 times the assessment of revenue.[25]

'Famine Commission of 1880 found that 1/3 of the land holding class in the country was in deep debt, while another 1/3 also in debt was in a position to repay it. An enquiry conducted by Thormburn in 474 widely distant villages in Punjab revealed that the cultivators in 126 villages were hopelessly involved in debt, in another 210 villages they were seriously involved and only in 138 villages the burden was comparatively light. The total indebtedness of 474 villages was estimated at Rs, 47.3 lakhs or roughly a little over Rs. 15 per acre.[26]

Dr. Bhatia concluded that the total debt of India in 1895 come to roughly Rs. 300 crores.[27] In 1901, the Famine Commission estimated that only less than 20 per cent of the cultivators were free from debt, and that 80 per cent was indebted to a greater or smaller extent.[28]

The crushing burden of state dues on one side and the uncertainties of agriculture on the other often obliged the small agriculturist to seek the help of the money-lender. Consequently, a strong trend towards transfer of land from the agriculturist to the non-agriculturist money-lender thereby separating agricultural capital from labour engaged in the cultivation of land developed in the second half of the 19th century. This had a very adverse effect on agricultural production. The cultivator himself lacked capital for investments in improvements. The moneyed classes found money-lending and agricultural property more profitable and were just not interested in agricultural progress. To sum up, the following factors contributed to the growth of debts.

(1) The growth of debt was due to the accumulation of

interest. The rates of interest charged were as high as 100 per cent in certain cases and varied from 18-40 per cent in most cases. The high rates made it difficult for the peasant to clear the interest charges alone, leave aside the principal.[29]

(2) Another factor was institution of individual ownership in land which could be mortgaged and transferred for payment of debt. By so doing, the authorities made an unconditional gift of a valuable asset and raised the credit of every peasant proprietor from the former limits of the surplus of an occasional goal crop to the market value of the proprietary right confirmed.

Increasing Food Exports

In order to meet the Home Charges, it was necessary for India to maintain an export surplus in her balance of trade. This she did by exporting her agricultural commodities including foodgrains. Food exports alone rose from 0.65 million tonnes in 1867-68 to 1.25 million tonnes in 1879-80. "Even during the five years ending 1894-95, exports averaged 1.25 million tonnes per year. Such a large exportation of Indian grains to England resulted in scarcity of foodgrain in India, sir James Cairol observed that "the principal cause which made the famine in Punjab in 1878 was the large exportation of wheat to England."[30]

Poverty

India was ruled, not in the interest of Indians, but in the interest of Britain. This resulted in the ever increasing poverty of the people which incapacitated them facing the recurring famines. The net per capita income in India has been variously calculated. Dadabhai Naoroji estimated at the end of the 19th century, the per capita income in British India as £ 2 while that of U.K. as £ 30 per person per annum,[31] and J. D. Rees made a rough estimates at rupees 30.[32] Rural indebtedness was also one of the causes of poverty of India, Sir Edward Maclagan estimated the debt burden at rupees 300 crores in 1911 and Mr. Darling at rupees 600 crores in 1928.[33]

The major factor responsible for the poverty of the rural population was unequal distribution of the main source of the rural income i.e. land. The distribution of land as noted in 1953-54 is presented in Table 1.5. About five per cent households owned 42.92 per cent of the total land the size of their holdings being 20 acres or above whereas 47.26 per cent households had either no land or had holdings upto only 1 acre, and they owned only 1.37 per cent of total area. Nearly 12.77 per cent households had holdings of 10 acres or more and the area owned by them accounted for more than 65 per cent of the land. Had land not been so unequally distributed among the landholders, poverty of the rural population would have not been so alarming and widespread.[34]

Table 1.5
Percentage Distribution of Households and Area Owned by Size of Ownership Landholdings (1953-54)

Size of land-holding (acre)	*Households (per cent)*	*Area (per cent)*
0.00 (landless)	23.09	0.00
0.1 -1.0	24.17	1.37
1.0 - 2.5	23.98	4.86
2.5 - 5.0	13.49	10.09
5.0 - 10.0	12.50	18.40
10.0 -20.0	7.71	22.36
Above 20	5.06	42.92

Source: Computed from N.S.S., 8th Round (1953-54) July 1954 to April 1955.

Nearly 50 per cent of the total urban population seems to have remained poor during the British period. If we take consumption level as an indicator of poverty, all those who had consumption below Rs. 15 per month i.e. 50 paise per day may be regarded as poor considering the cost of living indicators. Such urban population as had consumption below

Rs. 15 per month was 38 per cent and about 50 per cent of the urban population consumed less than Rs. 18 per month. Thus we arrive at a conclusion that about 40to 45 per cent of the urban population was poor during the British period.[35]

Bengal Famine

India experienced a number of scarcities but no major famine over a period of thirty five years from 1908 to 1942. The scarcities did not involve considerable loss of life. It was in 1943 that Bengal was confronted with severe famine.

There was a shortage -in the supply of rice in 1943. The main reason for this was the low yield of the *aman* crop reaped at the close of 1942. Another reason was that the stocks carried over from the previous year i.e. 1942, were also short.

The shortfall in foodgrains supply in Bengal was to the extent of about 10 per cent of the requirements. However, coupled with the stoppage of importation of rice from Burma and the dislocation of normal channels of trade due to hostilities at the border, prices skyrocketed. The Bengal Government failed to secure control over rice supply and distribution. Consequently, price rise was upto 4 to 5 times the normal prices. This price rise viewed in relation to the poverty in Bengal explains the vast death-toll.[36] Even in normal years about 5.5 million families afforded only bare subsistence. Lacking money to buy rice, people moved to Calcutta and died by thousands in Calcutta streets, while god owns were full of rice. The famine has therefore, been called a man-made famine. Official estimates put the figure of deaths at 1.5 million while unofficial estimates of deaths range between 3 and 4 million people.[37]

Millions died despite the fact that,' "Food of all sorts lay before their eyes, arranged in heaps, in shops all over the city. But no one attempted to seize it by force. ... The attitude of people themselves was that of complete resignation. The uncomplaining surrender to death by starving was the Bengali adaptation to famine. In European tradition, famine violence was turned "outward" and "upward" against landlords, merchants and officials. In Bengal the tradition was to turn violence "inwards" and "downward" against clients and dependents....Mendicancy, cries and wails, imploring gestures...all were part of the destitute attempt to evoke charity...when begging proved inadequate for subsistence, "fatalism" ensued,"[38]

The Bengal famine brought forward clearly the economic causes of famine in India. With progressive impoverishment of the people during the British rule about 40 per cent of the population lived below the poverty line. As a result even with marginal shortage, the situation turned, critical. The remedy lay in the overall improvement in economic conditions of the poorer sections of the population and in the increase in foodgrains production. The latter was achieved in the post-independence period. This together with the improvement in the transport and communication network and the end of the foreign rule led to a basic change in the situation. Drought did not result in large scale deaths. However, the problem of massive poverty due to lack of employment and low wages persists. Hence, the need for drought relief persists in post-independence India.

References

1. W. Digby (1961), *Prosperous British* India, p. 166.
2. K. Maloo (1987), *The History of Famines in Rajputana,* p. 5.
3. R. Datta, *Economic History of India,* p. 278.
4. H. Srivastava, *The History of Indian Famines,* p. 101.
5. H. Mukherji, *Economic History of India,* p. 5.
6. K. Maloo, *The History of Famines in Rajputana.*
7. H. Srivastava, op. cit.
8. *Ibid.*
9. T. Singh (1978), *D.P.A.P. in India,* p. 1.
10. B. M. Bhátia (1962), *Famines in India,* p. 239.
11. *Ibid.,* p. 240.
12. *Ibid.*

13. *Ibid.*, p. 34.
14. M. M. Malviyas (1979). Note of dissent Indian Industrial Commission Report, Quoted in G. Kaushala, *History of India*, p. 24.
15. C. N. Vakil (1924). *Financial Development in India*, p. 372.
16. I.F.C. Report of 1880. Appendix I, p. 143.
17. B. Smiths. Report of 14th August 1861, para 36, given in Srivastava p. 351.
18. R. C. Dutta, Famines and Land Assessment in India, p. 17.
19. R. C. Dutta, *Economic Victorian Age*, p. 470.
20. *Ibid.*, p. 496.
21. *Ibid.*, pp. 464, 467.
22. Baird Smith, Report, given in Srivastava, p. 353.
23. H. Srivastava, *op. cit.*, p. 351.
24. *Ibid.*, p. 355.
25. Deccan Riot Commission Report 1878, p. 29.
26. B. Bhatia, *op. cit.*, p. 151.
27. *Ibid.*
28. I.F.C. Report of 1901, p. 334.
29. G. Kaushal, *History of India*.
30. P. C. Ray. (1901). *Indian Famines: Their Causes and Remedies*, p. 19.
31. K. Dadabhai *op. cit.*, p. 242.
32. J. D. Rees. *Famines in India*, p. 283. Quoted by Loveday, p. 112.
33. Jathar and Beri (1939), *Indian Economics*, p. 283.
34. D. Dwivedi (1971), *Economic Concentration and Poverty in India*, p. 42.
35. *Ibid.*, p. 44.
36. S. Brahme (1983), Drought in Maharashtra, 1972, p. 8.
37. Aykroyd (1974). *The Conquest of Famine*, p. 77.
38. Greenough (1982), *Prosperity and Misery in Modern Bengal: The Famine of 1943-1944*, p. 266.

2

Economic Impact of Famine

Economic effects of famine are well known. It disturbs the prevailing order of production, increases unemployment, reduces purchasing power, and creates serious food problems. The immediate effect of crop failures is not only to reduce the food availability in the affected region, but also to shatter the rural economy.[1]

The economic impact of famine during the British rule can be summed up as follows. With the failure of rains the landless agricultural labourers found little employment as farm activity was brought to a standstill while general impoverishment simultaneously enlarged the supply of casual labour. Food prices increased as the less vulnerable groups strived to maintain a reasonable food consumption level, while trade was often slow to move food to the affected area from other regions. Wages to lagged behind the increased price levels.[2] Further aggravating the plight of agricultural labourers, the operation of the so-called "moral economy" did little to mitigate their sufferings which all too often ended only in death. Thus, severe famines frequently took place even when crop failures were only localized and food was far from wanting in the country as a whole.

This recurring scenario was aptly summarized by Baird Smiths' well known statement to the effect that famines in

India are "rather famines of work than of food."[3] The same verdict was arrived at later by successive Famine Enquiry Commission as well as by roost independent analysts though there has been much more controversy about the persistent causes of poverty in the same period. The Famine Commission Report of 1880 is worth quoting. "The first effect of drought is to diminish greatly and at last to stop all field labour and to throw out of employment the great mass of people who live on the wages of such labour."[4]

"...distress is mainly among the agricultural portion of the population thrown out of work by the failure of their ordinary employment. And the few small trades and handicrafts which are chiefly dependent upon them for sale of their manufactures,...among this class, distress arises not so much from an actual want of food, as from a loss of wages, in other words money to buy food ... as a general rule there is an abundance of food procurable even in the worst districts...they starve, not from the impossibility of getting food, but for want of the necessary to buy it."[5]

Impact on Farm Income

The past and recent droughts indicated that they have a wide ranging impact on agricultural income, rural demand for agricultural and industrial goods and manufacturing output. The fall in agricultural production triggers a host of problems ranging from upset of the profile of income distribution to widening the trade gap due to import of essential commodities. The erosion of real income in view of an estimated increase in the consumer index by l4-15 per cent causes serious demand constraints for agricultural and manufacturing industries, according to a study prepared by F.I.T.C.C.I. titled "Economic Impact of Drought."

The study endeavours to quantify the impact of drought using certain economic indicators. Based on a model which envisages three scenarios of fall in agricultural production i.e. 10 per cent, 15 per cent and 20 per cent the study estimates the fall in agricultural in come to be of the order of Rs. 8,400 crores for 10 per cent fall, Rs. 13,600 crores for 10 per cent fall, Rs. 16,800 crores for 20 per cent fall. The fall in rural demand for the year 1987-88 on account of drought at 1985-86 prices was estimated to be Rs. 7,022 crores, Rs. 9,279 crores and Rs. 14,043 crores for 10 per cent, 15 per cent and 20 per cent fall in agricultural production respectively.[6] The study cautions that unless urgent remedial steps are taken 25 per cent of the existing small farmers and two third of landless labourers would go below the poverty line.

Rural demand for industrial goods is quite significant; for every Rs, 100 of agricultural income, expenditure on industrial goods may be estimated to be Rs. 24. In terms of 1985-86 prices, the fall in demand for industrial goods on account of the drought in 1987-88 can be anywhere between Rs. 2,023 crores to Rs. 4,045 crores depending on the extent of shortfall which is estimated to be between 10 and 20 per cent.[7] The study points out that the reduction, in agricultural income would adversely affect the off-take of items such as cotton, electronics, tractors, leather footwear, diesel engines, power driven pumps, electric fans, radios, etc.

Another disturbing feature of drought is its impact on agro-based manufacturing output. The study indicates that the shortfall in agro-based manufacturing output could be about Rs. 2,113 crores for 10 per cent fall in agricultural production, Rs. 3,169 crores for 15 per cent fall and Rs. 5,045 crores for 20 per cent fall at 1985-86 prices.[8]

The study observes that whether it is on account of shortages of agricultural raw materials or on account of fall in demand the impact of drought will be progressively felt during 1987-88 and more in the next year.

The study points out that the 1965 drought led to a fall in industrial production of 0.4 per cent in 1966 and 1.3 per cent in 1967. The fall in industrial production in 1987-88 due to the present drought is estimated to be between 1.3 and 8.5 per cent.

Another problem area is the trade deficit. With the sharp reduction in the production of edible oils coupled with heavy import of cotton, sugar, viscose staple and petroleum products, the import will get inflated. At the same time surplus available for exports will be drastically reduced.

Impact on Employment

One of the objectives of Indian planning is the progressive reduction of unemployment in the country. Although policies for reducing the instability in production have favourable pay offs, their impact is felt essentially in the long run. This is because the area subjected to drought in India is very large. Therefore, in the medium run, public employment programmes must form a part of overall drought relief strategy. The important questions that deserve analysis are the scale of works programmes, and their cost effectiveness. The responsiveness of employment to variations in output differs across the regions and as between different crops, depending upon the degree of labour saving technology adopted. In general this responsiveness may be expected to be lower in the green revolution states or regions such as Punjab and Haryana as compared to other regions. For all crops and for the country as a whole employment elasticity for the recent period is estimated to be 0.59.[9]

According to the G. Saibaba estimates of employment and unemployment for different ages of males and females and for rural and urban areas are available on the basis of the three concepts viz, usual status, weekly status and daily status. For our purpose the weekly and more so daily status are highly relevant as these estimates bring out the seasonal and part time unemployment and under unemployment. The data are given in Table 2.1. It would seem that out of the 12.02 million unemployed persons as on March 1980, 4.19 million were having sorce gainful subsidiary work. Further, analysis by age group showed that 3/4 of the unemployed under the usual status category were concentrated in the fresh entrants to the labour force. A significant finding that emerges from the analysis of National Sample Survey (N.S.S.) 32nd round (1977-78) is that the rural agricultural labour households covering about 21.2 per cent of the all India population account for 46.7 per cent of the total daily status unemployment. The unemployment rates were the highest in labour households. Self-employed households had the lowest rates of unemployment.[10]

Table 2.1
Estimated Unemployment in March 1980

(Million)

Concept	*Upto 15 year*	*15 year & above*	*15 to 59 year*
Usual status	12.02	11.42	11.31
Weekly status	12.19	11.64	11.36
Daily status	20.74	19.77	19.17

Source: Sixth Five Year Plan (1980-85), P. 204.

However, unemployment data is not available separately for drought prone areas. During 1986-87 it is expected that the shortfall in foodgrains output may be of the order of 13.5 per cent of the previous years output. The loss in-employment/income in a drought year such as the current, would then be about 8 per cent of the previous years. 'This of course, is an average for the country as a whole. In some regions such as Gujarat or Rajasthan where the output loss is expected to be nearly 50 per cent employment/income loss could be substantial (nearly 30 per cent). Thus how much employment should expand depends on the intensity of drought which differs from region to region. According to the estimates derived by A. de Tanvry and K. Subbarao for the year 1978, "nearly 80 per cent of gross value added by labour was contributed by the landless, submarginal and marginal holdings."[11] Thus the burden of the loss in employment in a drought year falls primarily on the households below the poverty line.

Effect on Level of Living

Due to the loss of employment there is a reduction in

the wage incomes of the agricultural labourers and small cultivators. It was experienced in 1972 drought in Maharashtra that if all the working hands in the family joined the relief work the wages earned, by and large, made up for the earnings normally obtained through casual agricultural work. Howèver, due to the rise in prices real incomes recorded a decline. The loss of income sustained by the medium and large cultivators was substantial because of failure of crop. The level of living was noted to be below the normal level of living by about 40 per cent for those who were primarily dependent on the scarcity relief work for their livelihood.[12] In the case of medium cultivators the reduction was to the extent of 50 to 60 per cent. If consumption of cereals alone is taken into consideration the proportion of population with cereal consumption below 12 kg per month per person increased from 38 per cent in normal period (1967-68) to 60 per cent in 1972-73.[13] This proportion was even larger in villages where the crop failure was total and the supply of cereals through fair price shops vas irregular and meagre.

Apart from the impact on the level of living experienced daring the drought year itself, there is a long term impact of drought in terms of loss of livestock, loss of assets, increase in debt etc. For example, in the case of the drought of 1972 in Maharashtra, it was reported that the loss on account of sale or death of livestock was as high as 37 per cent and 52 per cent respectively in the severely drought affected taluks of Ahmadnagar and Osmanabad districts. For the entire drought affected tract the estimated loss of cattle came to 13 per cent, the incidence being highest in the case of cows viz. 25 per cent.[14]

As to the sale of assets, the labour families and small cultivators have hardly any assets to fall back upon during the drought period. The medium cultivators with some assets in terms of land, cattle, gold etc, mortgage or sell some of the assets to raise ready cash to meet the daily necessities. In the study of drought in Maharashtra (1972) it was noted that about 23 per cent families reported reduction in assets.[15]

The cultivators who barely make a subsistence in normal

times are compelled in a year of crop failure to borrow, as the cash earnings from relief works hardly meet the basic consumption need. The drought affected families have thus to bear the adverse effects of drought for some years because of: (a) the increased debt burden due to new borrowings taken partly to meet the consumption needs, partly to make up for the lost assets (particularly draught cattle), and debt repayment rescheduling, (b) lack of funds for effecting repair, improvements etc. in cultivation. This means a lowering of level of living for a few years after the drought year.

Health and Nutrition

Droughts have a serious impact on the health of the population of the region. It reduces the life span and productivity of human beings. Lack of accurate perception of the drought hazard has brought innumerable miseries to the farmers of drought prone areas particularly in the desert fringe villages. Scarcity of water, grass and grain has become so much a part of drought prone areas that they excite no comment until these calamities are felt simultaneously.

An investment on health is investment on man and improving the quality of his life. Health has to be viewed in its totality as a part of the strategy of human resource development. Horizontal and vertical linkage have to be established among all the inter related programme like protected water supply, environmental sanitation and hygiene, nutrition, education, family planning and maternity and child welfare. Only with such linkage the benefits of various programme,' can be optimized. An attack on the problem of disease cannot be entirely successful unless it is accompanied by an attack unemployment and poverty itself which is the main cause of it.

The problem of malnutrition is widely prevalent across the various socio-economic groups particularly among those below the poverty line. Landless agricultural labourers, people in tribal areas and those who are affected by constant calamities like drought are more vulnerable to this phenomenon. Children, pregnant women, and nursing mothers are seriously affected by malnutrition and the damage they sustain could be irreversible.

A survey conducted by the National Institute of Nutrition (Hyderabad) in February 1973 in the worst affected talukas of Poona, Ahmednagar, Bhir and Aurangabad districts of Maharashtra, noted that the incidence of various deficiency signs observed were similar to those frequently seen among the poorest sections of the rural population in many other parts of the country.

Scarcity of Drinking Water

Due to the drought an acute scarcity of drinking water is faced in many parts of the country, as wells, tanks and other sources dry up in large numbers. Even under normal rainfall, one of the most serious problem which rural communities face concerns the supply of drinking water.[16] Of the 3.81 lakh villages in India 37,017 villages did not have a single source of drinking water and more than half of the total villages had only limited drinking water facilities i.e. only one source of water supply in early eighties. During the Sixth Five Year Plan (1985-90) a massive programme for providing drinking water facilities was introduced under the Minimum Needs Programme. However, during the drought year 1986-87 drinking water and fodder shortage was widespread. Acute scarcity of drinking water supply was experienced in 10,402 villages in Gujarat, 13000 villages in Maharashtra.

References

1. Jean Breeze (1988), *Famine Prevention in India,* p. 7.
2. B. Bhatia (1967), *Famines in India,* Asia Publishing House, Bombay.
3. Baird Smith. Quoted in Srivastava, p. 53.
4. Indian Famine Commission Report of 1880, p. 35.
5. Indian Famine Commission Report of 1880, Appendix I, pp. 205-206.

6. G.O.I. *Economic Survey of India,* 1937-88.
7. *Economic Times,* February 1988.
8. *Journal of Facts for You.* June, 1988, p. 85.
9. Shiela Bhalla (1987). *Trends in Employment in Indian Agriculture,* given in Hanumantrao, p. 89.
10. Journal of Khadigramodyog, February 1986.
11. K. Subbarao (1986), Given in Hanumantrao, p. 90.
12. S. Brahme,(1983), *Brought in Maharashtra 1972: A Case for Irrigation Planning,* G.I.P.E., Pune, p. 83.
13. *Ibid.,* p. 84.
14. *Ibid.,* p. 78.
15. *Ibid.,* p. 72.
16. Kurukshetra, July 1986, p. 22 and G.O.I. Sixth Five Year Plan 1980-85, p. 204.

3

Efforts for Drought Mitigation in Post Independence Period

During the period 1850-1900 famine visited India 17 times and the death toll taken by famines was about 20 million people. In addition there was a huge of loss of cattle wealth. Agriculture was ruined and poverty depened due to frequent famines. In the period 1901-1947 there was only one severe famine, the Bengal famine of 1943 that took a death-toll of 30 lacs.

After 1950 efforts were made through extension of area under cultivation and development of irrigation to increase foodgrains production. Yet the production was not sufficient and imports of foodgrains continued to be a normal feature till the seventies.

The dangers of dependency on imports for procuring foodgrains became clear during the 1965-66 famine. "The famine also revealed the dimensions of our poverty, the retrograde character of our lane; structure, the continued exploitation of our landless and small farmers."

How far the policies adopted in the post independence period to fight famine have been effective is reviewed in the following.

Fluctuations in Foodgrains Production

Indian agriculture is usually described as a gamble in monsoon. There used to be a great deal of uncertainty about crop prospects as monsoon played a decisive role in determining agricultural output and failure of rains used to result in widespread famines. In the post-independence period there has been rapid development of irrigation facilities and spread of new technology in agriculture. Yet the performance in relation to coarse grains, pulses and oil seeds has not been satisfactory. During the period 1950-51 to 1985-B6 the share of coarse grains declined from 30.3 per cent to 17.6 per cent that of pulses from 16.5 per cent to 8.6 per cent.[1] The area under coarse grains declined from 39 per cent to 31 per cent and the area under pulses from 20 to 18 per cent, while the area under wheat increased from 10 per cent in 1950-51 to 19 per cent by 1984-35. The share of rice in the total foodgrain production has remained stagnant at around 40 to 42 per cent while that of wheat increased from 12.7 per cent in 1950-51 to 31.2 per cent in 1985-36.[2]

There has been a fall in the production of coarse grains from 34.35 million tonnes (1983-84) to 29 million tonnes (1986-87). During the last three decades production of pulses has remained more or less stagnant, ranging between 10 and 12 million tonnes covering an area about 23.5 million hectares.[3] By the end of this century, the country would require 24.7 million tonnes of pulses for a population of 935 million.[4]

The position of oilseeds is not better. The country is experiencing edible oil shortage. The demand for edible oil was 5.14 million tonnes in 1986-87 while or the actual availability was 3.5 million tonnes.[5]

Fluctuations in output are not necessarily reduced with growth in output. Much depends on the pattern of growth and the agro-climatic and infrastructures environment within which growth is brought about. Use of modern inputs under conditions of assured rainfall or irrigation can result in smoothening the growth curve by reducing fluctuations whereas the application of the same technology may result in accentuating fluctuations around the growth path.

Table 3.1 shows that the amplitude of fluctuations for all categories of crops except wheat have increased in the post green revolution period (1968-1985) compared to the pre-green revolution period (1950-1965). The increase in variability is the highest in the case of oilseeds followed by coarse cereals and pulses which are grown largely under unirrigated and uncertain rainfall conditions.

Table 3.1
Instability in Crop Production—All India (Standard Deviation in Annual Output Growth Rate)

Crop and crop groups	*1950-65*	*1966-85*	*1968-85*
Rice	9.99	13.92	14.27
Wheat	11.52	11.15	11.44
Coarse cereals	8.01	13.20	13.35
Cereals	7.83	10.81	11.10
Pulses	12.00	16.26	16.07
Kharif foodgrains	-	-	13.88
Rabi foodgrains	-	-	9.83
Total foodgrains	8.10	11.40	11. 43
Oil seeds	9.30	16.92	17.36
All crops	6.27	9.19	9.40

Source: C. H. Hanumantha Rao, p. 20.

Role of Irrigation in Stabilising Production

Agriculture may be deemed to have been drought protected by irrigation if output under irrigated conditions remains on its trend course in spite of a drought. The effect of drought may be assessed by computing the per centage fall in irrigated crop output from its trend path. This can be contrasted against the corresponding decline in the rainfed output. Finally the elasticity of total output decline can be computed with respect to reduction in normal rainfall.

According to this criterion, no single state has achieved full drought, proof condition through irrigation. In some drought years, signs of drought proof situation are visible in the states of Punjab (1979-80), Rajasthan (1974-75) and Karnataka (1972-73). In 6 out of 11 states the evidence suggests that, irrigation has offered partial protection in the sense that the extent of decline in irrigated output has been less than the corresponding decline in unirrigated output during droughts. These states are: U.P., Rajasthan, Maharashtra, Karnataka and Gujarat. In Tamil Nadu, H.P., and Bihar the protective role of irrigation is obvious only during some of the drought years while in the cass of A.P. it was noted that reduction in output in irrigated segment during the droughts of 1972-73 and 1979-80 surpassed corresponding decline in the rainfed sector.

Taking all the states together the protective role of irrigation is clear under the conditions of a severe drought. Reduction in irrigated output, during the drought 1972-73 vas only 7 per cent below the trend level in contrast with a 20 per cent reduction in the unirrigate segment (Table 3.2). In the absence of irrigation the impact of drought, or agricultural output tends to be more in the low rainfall regions. The natural magnitude of the output elasticity with respect to rainfall was about 1.6 in the low rainfall states, 1.0 in the medium rainfall states, and about 0.6 in the high rainfall states. In the presence of irrigation these magnitudes become 0.2, 0.5 and 1.0 respectively. This suggests that the protective benefit of irrigation even though partial tends to be confined primarily to low and medium rainfall areas.

Exploitation of Ground-Water

While development of irrigation helped growth of agricultural production, the over-exploitation of ground-water resources is creating famine of drinking water.

Since the beginning of planning of India schemes for exploitation of ground-water schemes are supported by the government. According to the Dakshinamurti et al, "the very quick growth in the utilization of ground-water is based on

Table 3.2

Reduction in Output in Irrigated, and Non-irrigated Areas* in Selected Drought Years

Drought Year	% deficiency in rainfall	% reduction in				Elasticity w.r.t. rainfall			
		Total Output		Total food-grains		Total Output		Total food-grains Output	
		IR	UR	IR	UR	IR	UR	IR	UR
1972-73	27	7	20	8	14	0.3	0.7	0.3	0.7
1974-75	19	7	13	10	18	0.4	0.7	0.5	0.7
1979-80	18	10	20	6	22	0.6	1.1	0.3	1.2

IR = Irrigated region
UR = Unirrigated region
* In eleven states viz. A.P. Bihar, Gujarat, Haryana, Karnataka, M.P., Mahsrashtra, Punjab, Rajasthan, Tamil Nadu, U.P.

Source: C. H. Hanumantha Rao, p.54

the financial support given to energized pumps." Table 3.3 gives the ststavize rates of ground-water exploitation in India.

According to the Dakshinamurti. "It is seen that from 1950 to 1966 the development of ground-water was about 2.5 per cent in linear basis, based on the area irrigated from ground-water resources during the year of 1950-51, the growth rates from 1960-61 to 1964-65 was 3.7 per cent.[6] It suddenly rose upto 19 per cent from 1964-65 to 1968-69. This sudden and high increase in growth rate has been due to the advent of high yielding crop varieties, mobilization of institutional resources for financing the programmes and stepping up of rural electrification."

Table 3.3
State-wise Projections for Development of Gross Area Irrigated from Ground-water Schemes

(Figures in thousand hectares)

State	*Gross irrigated area*	*Projections for future development of gross irrigated area*			
	1968-69	*1973-74*	*1978-79*	*1983-84*	*1988-89*
1	2	3	4	5	6
Andhra Predesh	622.0	890.34	1214.10	1618.80	2225.85
Assam (Mizoram, Arunachal Pradesh and Nagaland)	-	20.23	80.94	223.05	485.64
Bihar	569.00	809.40	1133.16	1618.80	2225.85
Gujarat	890.98	1173.63	1416.45	1578.33	1618.80
Haryana	547.15	728.40	890.34	971.28	1011.75
Mysore	352.89	566.58	849.87	1173.63	1618.80
Himachal Pradesh	2.02	4.05	10.11	20.23	40.47
Jammu & Kashmir	2.02	16.19	40.47	80.94	141.64
Madhya Pradesh	467.83	928.46	1133.16	1618.80	2225.85
Kerala	8.09	20.23	50.58	101.17	202.35
Maharashtra	902.07	1214.10	1556.92	1578.33	1618.80

1	2	3	4	5	6
Orissa	40.06	80.94	202.35	406.70	607.95
Punjab	1949.03	2428.20	2651.02	2751.96	2832.90
Rajasthan	1426.97	1618.80	1821.15	1942.56	2023.50
Tamil Nadu	951.85	1335.51	1659.27	1780.68	2821.15
Uttar Pradesh	4171.24	5807.40	7446.48	9085.57	10319.85
West Bengal	16.18	283.30	484.64	809.40	1214.10
Union Territories remaining states	55.04	80.94	101.17	121.40	141.64
All India	12994.44	17806.75	22663.18	27519.57	23375.99
Volume of water	8.5	11.25	14.35	17.37	20.45

Source: Report of the task Force on Ground-Water Resources (1972).

"Most ground-water utilization in India is from the shallow aquifer zone with depth less than 400-500 feet...With the shallow aquifers totally exhausted, the dug wells and hanks will not store water very long creating pseudo drought condition. The intensification of this pseudo drought is due to encouragement to ground-water based irrigation, sometimes with the declared objective of drought relief."[7]

In Maharashtra over exploitation of found-water through tubeveils sunk mainly by the rich farmers particularly to irrigate sugarcane banana, another heavy water consuming crop, is proving a drain on the ground-water resources. The depletion of ground-water has increased the problem villages with inadequate source of drinking water to a staggering figure of 23,000 villages during the Sixth Five Tear Plan.[8]

Depletion of Forests

Depletion of forest cover has both direct and indirect implications for drought deforestation has been one of the main causes of soil corrosion; soil degradation in turn has contributed to output fluctuations so that the states with high intensity of erosion problem i.e. (Maharashtra, Gujarat and

Karnataka) are also the states subjected to periodic drought. The direct effect of soil erosion is the decline in the capacity to hold water which accentuates the vulnerable by of that region to drought. As tree cover disappears opportunity for recharge of surface flow and ground-water wre reduced. Possibilities for internal drainage of excess water in high rainfall regions also get reduced.

Valdiya reported that in the Ganga river catchments in Nainital district about 45 per cent of the springs have gone dry or become seasonal. Studies carried out at experiments watershed near Chandigarh reveal that burning of forests increases the peak discharge of run off by 69 per cent, the felling of trees by 34 per cent and overgrazing by 32 per cent.[9]

Data for selected storms of some agricultural and forest watersheds at Dehra Dun show that the forest watershed had about 10 per cent lower peak flow than the agricultural watershed. The forest watershed, gave 38.5 per cent less soil loss than the agricultural watershed. Another study showed a 28 per cent, reduction in run off and 73 per cent reduction in peak flow due to a forestation.[11] Experimental run off plots at Nunpur (Himachal Pradesh) shoved that the soil loss in cho area was 22443 kg/ha for grazed plots as against 306 kg/hectare from grass and shrub plot for one storm of 24.89 cm in 19 hrs on a 1 in 4 shope. The run off from the areas regularly grazed was 95 Per cent of the total rain compared to 20 per cent from the areas under shrub and grass cover.[12]

The rate of forest cover depletion is quite high in India (Table 3.4).

Apart from Madhya Pradesh where the loss is undoubtedly greatest, forests in the Western Ghats, spread over Karnataka, Kerala and Maharashtra have also been depleted considerably. Significant losses have also occurred in U.P, and West Bengal.

The details of the different purposes for which the forest area is lost are not, readily available at the all India level. However, the case study of Karnataka shows that "the major causes have been extension of cultivation, hydro electric projects, submersion and mining."[13] (Table 3.5). In a **study** of four villages in Karnataka, it is reporter that as much as 73 per cent of the encroachment on forest lands was done by the middle and rich peasants.[14] To the extent large and middle farmers owned a substantial proportion of livestock the benefits of free grazing on forest lands occurred to them proportionately more than to poor families.

Table 3.4
State-wise Forest Area Lost (1952-1983)

State	*Forest area lost (000 ha)*	*Area lost as % of geographical area*	*Area under forest as% of total geographical area (1982)*
A.P.	205.4	0.7	22.4
Assam	80.0	0.8	25.3
Bihar	70.0	0.4	16.6
Gujarat	188.6	1.0	10.0
Haryana	22.1	0.5	3.0
H.P.	32.9	0.6	14.6
J & K	90.9	0.4	13.1
Karnataka	311.3	1.6	15.8
Kerala	204.5	5.3	27.8
M.P.	1896.2	4.3	31.7
Maharashtra	221.8	0.7	17.3
Orissa	134.8	0.9	42.6
Punjab	11.5	0.2	4.4
Rajasthan	87.1	0.3	6.1
Tamil Nadu	67.7	0.5	15.6
Tripura	52.5	-	-
U.P.	237.1	0.8	17.4
Bengal	321.2	3.7	13.3

Source: (1) Forest Commission and (2) Statistical Outline of India 1986-87.

Table 3.5
Forest Ares Lost Since 1956 in Karnataka
(Purpose-wise)

Purpose	*Area cost in hectares*	*% of total area lost*
Hydroelectric projects (Kali, Chakra, Varahi etc.)	41068	18.4
Direct submersion	35840	16.1
Rehabilitation of the displaced	25820	11.6
Power lines	1688	0.8
Colony roads and townships	2121	0.9
Mining	42676	19.2
Other non-agri. use	6297	2.8
Extension of cultivation	67217	30.2
Total	222727	100.0

Source: Government of Karnataka, Gazetteer given in Hanumantha Rao, p. 130.

It will be thus clear from the above discussion that even though the extension of irrigation facilities has helped growth of foodgrains production, at the same time problems like over-exploitation of ground-water, deforestation and loss of forests due to submergence etc. are resulting in shortage of water supply. This in the long run might adversely affect the agricultural production. Though this problem is recognized concerted efforts to mitigate the adverse impact are not yet made.

References

1. Jean Dreeze (1988), *Famine Prevention in India,* p. 7.
2. *Journal of Facts for You,* June 1988.
3. *Ibid.*

4. *Ibid.*
5. *Ibid.*
6. *Economic and Political Weekly*, Vol. XXII, No. 50, 12 Dec. 1987.
7. *Ibid.*, p. 2165.
8. *Ibid.*, p. 2166.
9. C. H. Hanumantarao, S. K. Ray and K. Subba Rao (1988). *Unstable Agriculture and Droughts*, Vikas Publishing House Pvt. Ltd., New Delhi, p. 131.
10. R. C. Ghosh, B. K. Suubba Rao, B.C. Ramola (1980). *Interception Studies in Sal* (Shorea Rebusta) Cappice Forest in Indian Forester, 106(8).
11. H. N. Mathur, Ram Babu P. Joshie, Bakshish Singh (1076), *Effect of Clear Failing and Re-forestation on Run off and Peak Rates in Small Watershed*, Indian Forester 10(4).
12. R. L. Singh (1975). *The Problem of Silt in Relation to Irrigation and River Projects and Forests, Proceedings, Fourth Irrigation and Power Seminar*, Hirakund.
13. Hanumantrao, *op.cit*, p. 130.
14. *Ibid.*, p. 129.

4

Famine Relief in India

This chapter is divided into three parts. Part I deals with early relief efforts. Part II describes the drought relief since Independence, And Part III discusses measures and new approaches in for drought relief.

EARLY RELIEF EFFORTS

Famine relief has a very long history in India. One of the first treatises "Artha Shastra" written more than two thousand years ago and attributed to Kautilya, pronounces that when famine threatens, a good king should "Institute the building of food or water works with the grant of food or share provisions or entrust the country."[1]

According to Shrivastava. "The chief methods of famine relief adopted by Indian rulers included free distribution of raw grains, opening of free kitchens, opening of public grain stores to the people, remission of revenue, payment of advances, remission of other taxes, construction of public works, canals and embankments, sinking of wells encouragement of migration and increase in the pay of soldiers. Even sold-children in the time of Shahajan are spited to have been ransomed by the government and restored to their parents.[3]

However, the evidence is too scanty to judge the real

efficiency of relief efforts prior to the 19th century, and it in not implausible that they were far from being systematic and comprehensive.[3]

Under the rule of East India Company, famines were frequent and severe. Sometimes extremely severe as with the calamitous famine of 1770 in 'Bengal. Relief efforts were moreover, at best half hearted, and in any case lacking in effectiveness. The laconic remarks of the Famine Commission of 1880 on this subject are revealing enough.

> "...the earlier despatched of the Bengal Government, while breathing a tone of sincere compassion for the sufferings occasioned by famine are busied rather with its fiscal results as affecting the responsibility of the company towards its shareholders, than with schemes, which would have seemed wholly visionary for counteracting the inevitable less of life.[4]

The Famine Commission report of 1880 contains a vivid account of the nature of relief efforts prior to the famine codes.

"What often happens now is that they wander from their village, crowd into towns, die about the roads, and otherwise attract the attention of the officials. Then a survey is made, a relief work is started, and then follows all, the train of difficulties attendant on the endeavour to get masses of wretched, demoralized, half-starved creatures to work and be paid after some sort of method. The work is generally started too late to save life numbers, from one cause or another, do not get within its scope, every department is strained to supply supervision, and the supervision is generally quite inadequate for anything like real control, the wage is a hopeless dilemma, if yon give a law rate, the people desert and die, if you give a high one, yon drain the labour market and the thing gets beyond control".[5]

Before the large scale development of the railways from the 1870s onwards, private trade in foodgrains within India notably lacked in dynamism, and local scarcities precipitated very damaging price hikes.[6] Often, the lack of satisfactory communications (....) severely restricted the movement of foodgrains so that while in one part of the country people died of lack of food, in another, only a few miles away there was an abundance of cheap food."[7]... while in one bazar, famine prices of Tour rupees per maund might be ruling, in another, not thirty, miles off, the price would be but about rupee one and half for the same quantity, yet no flow from the full to the exhausted market could take place, because roods were not in existence and means of carriage unknown."[8]

"The extension of railways, and the connection of trunk lines, has so increased the rapidity of communication that mercantile relations now subsist between the native traders or all parts of India, and these traders keep themselves veil posted up in the state of the most distant markets, being keenly alive to the advantage of the telegraphic communication (....). The combined effect of railway and telegraph extension throughout the length and breadth of India, has permitted government to rely upon the activity of private trade for the supply of foods to all districts immediately served by railways (...). Moreover, the area from which food supplies can be drawn has been extended from the limit of 100 miles, which with a cart carriage, in a famine striken country destitute of fodder for cattle or oxen with pack-bullocks or cooly labour, is a maximum to a range of over 2000 miles, (...) as certainly as a strong demand arises for grain or other country produce, either for a famine district or for export, the railway stations of all districts from which export is possible, are crowded with stores of grain, while the railway officials are besieged by applicants for early despatch of their consignments."[9]

The Famine Commission of 1898 pronounced that "one of the most remarkable features in the recent famine was the uniform level of prices all over the country which is attributable to the ever-extending system of railways and which, if it increase the area greatly diminished the intensity of distress."[10]

The sudden dynamism of private trade may have been exaggerated, by the famine commission anxious to rationalize the policy of non-interference. Etheridge aptly called it "supposed infallible laws of the great masters of economic science.[11]" While the expansion of the railways was effective in ensuring the advance of private trade, it is much less obvious how beneficial this development was to various sections of tie Indian people.

Analysts such as Love-day emphatically concurred with this view, arguing that "the desirability of a system by which the pros pert is should help to bear the burden of the distressed is unquestionable."[12] But there also existed another school of thought which stressed the evils of the railways in generalized its possible role in facilitating exports from famine striken areas in particular.

In short, under native rule, famine was an evil which wag unmanageable, government's financial resources were limited and administration far from efficient; means of transport primitive as they were could not enter fodderless and roadless country and broke down just where it was most required. That is why a famine was disastrous in its consequence. Even though the development of railways after 1870 helped movement of grain, it was left to private initiative. Morsover, expart of foodgrains was not controlled even it famine year. Shortage of grain led to starvation and pestilence leading to large number of deaths particularly when the famines were widespread.

As to the opening of relief works to make up partly for the loss of income due to loss of employment the policy was to frame measures in such a way so as to "avoid every tendency to relax in the people the sense of obligation which rests on them to provide *for their own support by their own labour,* to cultivate habits of thrift and forethought and as for as possible to employ the surplus of years of plenty to meet the wants of years of scarcity,"[13] (IFCR, I, 1880 p, 35).

The relief provided was to be the lowest amount sufficient for the purpose of maintenance so as to save life. There was strict, discipline and stringent work norms and the wage rate were low so as to deter the poor and seasonally unemployed from flocking to the relief works. With so much of ordinary poverty prevailing in the country, and so much of unemployment in agriculture, the government wan wary about opening of relief works. Large public works fully supervised and at a distance from the village were the types of works usually opened by the government, to keep the numbers employed at the minimum so that the relief expenditure remained low. As a result of such a stringency the relief offered was far below the relief needed and consequently there was much hardship and thousands of deaths due to famine.

DROUGHT RELIEF IN INDIA SINCE INDEPENDENCE

The peasants, artisans and service workers and agricultural labourers are affected to varying degrees due to drought. The first casualities of famine are the artisans and other service workers and agricultural labourers as drought reduces the number of days of agricultural employment, leading to a, fall in their wage earnings. The extent of decline in their incomes depends upon the provision of wage-work through relief works.

Relief Works

Large public works were considered to be the backbone of relief works till 1901 while in the present century the emphasis shifted to village level works. The large public works include extensions and improvements of the existing irrigation system, the construction of minor irrigation works, railway earthwork and collection of metal and road works. Small village works include minor irrigation works, digging of community veils and collection of metal and road works. During the drought of 1972 relief centres employing a large number of workers had to be opened and most of the works opened could be classified as village works. They were within 5 kms, of the village and did not necessitate opening up or camps, shelter and provision of related facilities. If the distribution of the expenditure on different types of relief works is examined for 1972-73, it WAS noted that about 30 per cent of the expenditure was accounted for by community wells and about 20 per cent by percolation tanks and desilting of village tanks etc. About 22 per cent of the expenditure was absorbed by earth work for road construction and repair, 14 per cent by metal collection and 11 per cent by soil conservation work.[14]

If the work is evaluated on the basis of its utility, about 45 per cent of the expenditure was of doubtful utility about, 35 per cent of temporary utility and 20 per cent of permanent utility. The share of works of permanent utility was low as the relief works no more included large public works. They were mostly village level works, which necessarily have a very limited scope and coverage. Moreover, despite the repeated emphasis laid by famine commissions on keeping a plan of action in preparedness to be put into effect under drought conditions such preparedness is totally lacking. Consequently works are opened up in an adhoc manner with priority' given to the types of works which do not call for any survey, designing etc.

Employment Programmes

These programmes began essentially as a part of the famine relief measures. Even the recent state level programme in Maharsshtra viz. Employment Guarantee Scheme (E.G.S.) originated essentially as an out growth of drought relief measures in drought prone regions in Maharashtra State, With the accumulation of surplus food stocks with the government since the late 1970s, two national level programmes were launched to serve both as anti-poverty interventions and also as employment support measures in years of crop failures.

These are National Rural Employment Programme (N.R.E.P.) and Rural Landless Employment Guarantee Programme (R.L.E.G.P.).

Past experience suggests that, employment programmes of this kind do perform the' important role of stabilizing income levels of poor households over time. For example, the role played by E.G.S. in Maharashtra in combating seasonal troughs in incomes is substantial. It was reported that cultivators in western Maharashtra flocked to E.G.S. sites after the sowing operations were disrupted by lengthy dry spells during 1979. E.G.S. attendance rose sharply and remained high in vidarbha when August flooding destroyed the paddy crop. Similarly, a poor rabi crop was reflected in unusually high E.G.S. attendance during April-June 1983.[15]

There is also some direct quantitative evidence on the role of works programmes in reducing income instability. For example, landless labour households that relied almost entirely on earnings in the daily agricultural labour in market in Shrirapur and Kanzara villages in the dry regions of Maharashtra where E.G.S. operated since 1977, had shout 50 per cent less variable income streams as compared to the incomes of those in Aurepalle village in the arid zone in Andhra Pradesh, where E.G.S. did not operate.[16]

In terms of absolute amount of employment generated, a declining trend is noticeable for N.R.E.P. (Table 4.1) R.L.E.G.P. has been in operation only since three and half years. During 1987 it is expected that N.R.E.P. and R.L.E.G.P. would together generate not more than 250 million and 236 million mandays of employment respectively in the country as a whole.[17]

The employment targets for the two programmes respectively in 1987 are 274 million manday and 254.8 million mandays. It would interesting to compare the public employment generated to the existing stock of employment in rural India, and to the incremental growth, of this employment. Thus the present, scale of employment via public works is generating roughly 60 per cent of incremental growth of employment in rural India.

Apart from the extent of employment the unit cost of employment generationate the wage offered and the food component of wages are the other important aspects of work programmes. The unit cost per manday, the food component and the implicit wage cost are shown state-wise in Table 4.2. The unit cost per manday of employment worked out to Rs.

Table 4.1

Employment Generated and Cost per Manday of Employment under N.R.E.P. and R.L.E.G.P

	1982	*1983*	*1984*	*1985*	*1986*	*1987*
National Rural Employment Programme						
Total employment generated (million mandays)	354.52	351.2	302.76	353.12	316.0	256.3
Total cash expenditure (Rs. million)	3184.8	3916.2	3928.9	5014.8	5300.0	3960.0
Quantity of foodgrain distributed (million metrictones)	0.233	0.172	0.147	0.171	0.581	0.780
Per manday cash expenditure (Rs.)	8.98	11.28	12.98	14.2	16.77	15.5
Cash expenditure in real terms (at 1980-81 prices)	8.26	10.12	10.65	11.21	11.46	11.0
Foodgraine component (kg)	0.66	0.49	0.49	0.48	1.84	3.1
Wage cost* (including value of food grains in Rs.)	5.63	7.74	8.16	8.92	11.6	14.2
Wage cost in real tennes (at 1980-81 prices in Rs.)	5.15	6.97	6.70	6.70	7.9	10.0
Material cost (as a percentage of total cost)	37.6	39.7	37.6	39.6	upto 50%	upto 50%

	1982	1983	1984	1985	1986	1987
Cost per manday* Cash expenditure foodgrain)	10.0	12.0	13.7	15.0	20.0	23.10
Cost per manday at 1980-81 prices	9.0	10.2	10.6	11.0	30.7	
Rural Landless Employment Guarantee Programme						
Total employment genenrated (million mandays)			5.196	255.2	231.88	175.13
Cash expenditure. (Rs. million			62.109	3717.263	4375.5	3536.7
Foodgrains (metrictones)			1483	108305	399000	442000
Per manday cash expenditure. (Rs.)			11.95	14.57	18.9	20.2
Cash expenditure at 1980-81 prices			9.28	10.64	12.92	-
Foodgrain component per manday (kg.)			0.285	0.42	1.72	2.53
Cost per manday* (cash+Foodgrain)			12.4	15.3	22.0	24.6
Cost at 1980-81 prices			9.6	11.2	15.0	-

* At current prices

Sources: Computed from CAARD Report. Annual Report of the Ministry of Rural Development 1985.

20.6 as per the quarterly progress report of the 20 point programme, June 1987. During 1986 the average for the country was Rs. 20 (according to previous table).

Table 4.2
Wages and Cost in Different States Under National Rural Employment Programme (June 1987)

(Per manday)

State	*Implicit wage rate Rs.*	*Unit cost Rs.*	*Per capita foodgrains Rs.*
Uttar Pradesh	10.54	20.27	3.76
Bihar	17.12	27.07	5.11
Madhya Pradesh	10.34	16.15	3.79
Andhra Pradesh	13.06	23.42	2.75
Maharashtra	9.34	18.94	0.33
West Bengal	12.18	21.12	3.37
Tamil Nadu	9.8	11.66	4.42
Rajasthan	7.3	13.67	1.26
Karnataka	9.0	22.43	1.02
Gujarat	14.1	23.76	3.46
Orissa	10.4	18.00	2.70
Kerala	9.7	16.5	1.91
Assam	12.4	21.47	0.74
Punjab	20.4	36.99	7.44
Haryana	21.47	61.83	4.5
Jammu & Kashmir	19.34	24.02	7.5
Himachal Pradesh	13.47	17.32	5.46
All India Average	12.92	23.21	3.5
Standard Deviation	4.08	11.15	2.07
Co-efficient variation %	31.60	45.00	54.00

Source: Implementation of 20 Point Programme Quarterly Progress Report, June 1937, Ministry of Programme Implementation.

There appears to be significant interstate variation. Unit cost is of course affected by the extent of material cost involved, which in turn is determined by the nature of works undertaken. Wages are paid partly in cash and partly in kind (foodgrain's); it would be useful to convert the later into cash wages and compute the implicit wage-rate per manday for each state. This again shows substantial interstate variation. The interesting point to note is that in the drought prone states of Maharashtra, Rajasthan and Karnataka, the food component per manday is very small whereas it is the highest in the Punjab. In drought prone regions food price generally goes up sharply in drought year especially if they are not served by an effective public distribution system.

This analysis is based on the assumption that the wage rates actually paid at the site are as per the norms prescribed by the government. But this was not the case in reality.[18] Wage rates paid were very much below the prescribed norms and there was wide variation even within a district. The averages wage cost per manday of employment aggregated for the country as a whole remained unchanged at Rs. 9 over the period 1983-86 which was lower than the minimum wage in most states and lower than the market wage in some states."[19]

In Assam agricultural labourers were earning Rs. 14 in the market while the minimum wage was about Rs. 11 per day but only Rs. 8 were offered by N.R.E.P. In Gujarat, it Is reported that two different N.R.E.P. wages (Rs. 4 and 6 per day) were prevalent in two different talukas of a district. For Gujarat State as a whole N.R.E.P. wages actually paid ranged between Rs. 4 and 9 per manday across talukas which were lower than both 20 minimum wages and market wages.[20] In Rajasthan, actual wages averaged Rs. 10 as against the prescribed minimum wage of Rs. 14. Irregularities in ware payment including long delays in payment have also been noted at many places.

Another important, point of controversy has been the policy of wage payment on a piece rate basis. This method has been resorted to essentially to monitor productivity of works programmes. It is well known that nutrition levels of labourers fall during drought years. The energy level of labourers in teams of earth moving capacity is likely to go down in the drought affected regions. Therefore, the norms prescribed for piece rates during normal years may have to be relaxed during drought years to allow for the possible decline in the energy levels of labourers. If normal year norms are insisted upon in drought years, labourers may. in order to compensate for the decline in the energy levels, have to draft other family numbers, especially children, for earth-moving work which would be inimical to child's health.

MEASURES FOR DROUGHT RELIEF

There are two aspects to the issue of drought: (1) A short term plan to provide relief to the affected people and create the necessary infrastructure that would confer long term benefits. (2) A long term plan that would help to often the rigors if drought conditions occurred because of the failure of the monsoons. The short term plan should be so evolved as to intensify relief works and provide employment in drought hit areas so that migration could be halted. Emphasis is put on extension of plan schema and rural works programme at the local level that include soil conservation, digging tanks, locating ground-water resources, drinking water, cattle preservation nutrition programme etc. Measures have also to be taken to expand public distribution systems, and to hold the price line and curb malpractices in distribution. Though if, has become a routine verbal exercise to stress that Till relief work should create assets, in practice most of the relief activities fail to achieve this objective.

On a long term basis, there is a need for coordination of all the programmes related to agriculture and water management. The execution of programmes to boost agricultural production calls for control and monitoring through decentralized administrative system. There should be delegation of powers to the agencies in the field. In fact, effective coordination between scientific bodies, official, and non-official agencies is required in the implementation of farm development activities. The ultimate goal, as resources become available, should be to merge drought relief which is generally a collection of ad hoc measures, with long term development programmes designed to improve the economy, in the drought prone regions and correct the regional imbalances, accentuated by the green revolution. In effect, the drought roller work should serve strengthen the content of development programme which in turn would help to meet the physical requirements of rain fed and dry land areas. In other words, the drought fighting programme should be based on both short term and long-term approaches as they are not mutually exclusive. The routine exercise of allocation of funds for fodder, food and water that is ritualistically followed every time there is a drought, should be replaced over a period of a time, with long term measures that would include the programmes for "expanding; irrigation potential, taping and utilizing ground-water mass-scale bunding, introduction of drought resistance and short duration crops, development of infrastructure in the traditionally drought prone areas creation of fodder banks, and foodgrains buffer stocks, in rural area. More than these comprehensive steps for the development of dry land farming technology should be promoted in a systematic manner. The introduction of crop insurance schemes is a welcome move, since it gives a sense of security to the farmers.

Drought Mitigation

20 per cent of the area in India is periodically subjected to onslaught of droughts, leading to huge losses of agricultural production and livestock wealth in addition to the loss of water and of power generation.[21] Huge amounts have been spent by the governments to provide relief after the occurrence of droughts, but apart from providing much needed relief at the appropriate time such expenditure has not helped solve the basic problem of increasing the productivity of

these areas and reducing thereby the impact of the severity of the droughts on the human and cattle population.

Development of those areas in a most challenging task and a modest attempt has been made by the government through the central sector scheme of "Drought Prone Areas Programmes" (D.P.A.P.) Although this programme has been in operation since the Fourth Plan (1969 to 1974), it has been taken separately as an integrated area development programme only, during the. Fifth and Sixth Five Year Plans.

D.P.A.P. covered 557 blocks spread over 74 districts in the country in 1985.[22] It is an integrated area programme in the agricultural sector that aims at optimum utilization of land, water and livestock resources reducing the severity of the impact of drought. Some of the important elements of the D.P.A.P. are as:

1. Development and management of water resources,
2. Soil and moisture conservation programme,
3. A forestation with special emphasis on social and farm forestry,
4. Development, of pasture lands and range management in conjunction with development of sheep husbandry,
5. Livestock and dairy development,
6. Destruction of cropping pattern and change,
7. Development of subsidiary occupations.

While this was the broad policy frame for planning and implementing the D.P.A.P during the Sixth Plan, the review of the programme revealed a large number of basic deficiencies particularly in respect of availability of data and management, supervision, maintenance of assets, budgetary provision and administration.

An attempt is therefore made here to briefly review the progress and highlight the basic deficiencies which need to be corrected,

During the Sixth Five Year Plan (1980-85) an expenditure of Rs. 315 crores was incurred under the programme.[23] Irrigation, forestry and pasture, soil conservation and animal husbandry accounted for about 85 per cent of the total expenditure. As regards the physical achievement of the programme during the first four year of the plan 3.86 lakh hectares have been treated for soil conservation, irrigation potential has been created for 4.94 lakh hectares and forest and pastures have been raised over 3.42 lakh hectares. Employment for about 147 million mandays had been generated.[24]

For assessing the financial returns to the investment, it is necessary to concert the above physical returns into income gains. For this purpose, the report of the evolution of D.P.A.P. In Gujarat is useful, considering that drought intensity is usually severe in this state. According to this report the incremental income gains per acre for the first three Investments were as follows.[25]

Item of Investment	*Incremental gains per acre at (1076 prices) Rs. Range*	*Average*
1. Minor irrigation	200-418	330
2. Soil conservation	170-440	290
3. Afforestation and pasture	-	50

By far the most sizable and definitive increase in income came from minor irrigation. Annual household incomes rose by 50 per cent to 75 per cent, as cropping intensity levels reached, 150 per cent and in some cases even 200 per cent.

However, in some of the backward villages where farmers did not adopt the new technology the return per acre was only Rs. 200. But in the advanced talukas where farmer adopted the new technology following irrigation facilities, the return from minor irrigation amounted to Rs. 418 per acre. It must he stressed that even though there are virtually no gestation lags for minor irrigation it is necessary to allow for some time so that input supplies improve and farmers shift to new technology and/or more profitable cropping patterns. There were wider variations in returns to soil conservation schemes, depending upon the physiographic conditions.

The basic objective of the programme has been restoration of ecological balance through scientific development of land, water, cattle and other resources. However, there has been inadequate focus on items/schemes/ projects which should haves formed the core of drought proofing measures.

A watershed is the most suitable unit of planning for optimum conservation and development of land water, livestock resources. The schemes prepared by the department lack in respect of scientific planning on watershed basis.

It is observed that, various components of the are through the technical department of the state government. The he-ails of those departments at the state level have however, not been providing adequate guidance, technical supervision while planning and implementing these programmes.

A number of schemes are implemented without proper appraisal and without taking Into consideration the cost benefit, analysis as well as backward and forward linkages of the schemes. It is also observed from the review exercise that a number of states have not made adequate budet provision.

To mitigate the adverse impact of weather cycle it is essential to take the following measures.

1. Minimise the amplitude of fluctuation through protective irrigation through watershed planning in the drought prone areas. The criteria for use of water in these regions should generally be maximization of the returns in terms of crop production to the unit of water used. Propagation of scientific method of water preservation and application to help moisture and reduce risk is necessary.
2. Shift to livestock based activities and non-agricultural activities in the critical scarcity ridden areas which cannot be covered with protective irrigation.
3. Utilize the available water resources in other areas to increase the yields of foodgrains and essential crops.[26]

4. Place groundwater utilization on a scientific footing. In India there is little control over groundwater exploitation. There is an urgent need to evolve a scientific water resource policy that plans conjunctive use of surface water and groundwater and takes care that the ground-water resources are not depleted.

To enforce ecological water resource policy, the people of India must regain the lost control over their water resources. New awareness through experiments like Pani Chetana in Rajasthan, Pani Panchayat or Mukti Sangarsh in Maharashtra, Ganga Mukti in Bihar and many more both in the voluntary action sphere and in the various research organizations will bring the collective wisdom for realizing the ecological land and water resource policy. The main components of the ecologically sound policy are (i) conserve maximum quantum of rainfall water *in situ;* (ii) revegetate the catchments areas of water source and ban free grazing (iii) ban commercial and indiscriminate cutting of trees and promote greening of the catchments area, (iv) treat all land and water resources as social resources and plan "their utilization according to capability. To implement the scientific resource use plan, it is essential to devise appropriate institutional and fiscal policies.[27]

References

1. *Arthasastra,* Quoted in Chetty & Ratha (1987). p. 3.
2. H. Srivastawa (1968). *The History of Indian Famines* (1858-1918), p. 28.
3. Jean Dreeze (1988). *Famine Prevention in India,* p. 13.
4. I.F.C.R. of 1880, p. 3.
5. I.F.C.R. of 1880, Appendix I, p. 113.
6. Mc Alpin (1983). *Subject to Famine, Food Crises and Economic Change in Western India* (1860-1920).
7. *Ibid.*
8. Baird Smith. *Quoted in Srivastava,* p. 53.

9. I.F.C.R. of 1880, Appendix I, pp. 198-199.

10. I.F.C.R. of 1898, p. 351.

11. A. T. Etheridge (1963). *Report on the Past Famines in the Bombay Presidency*, p. 17.

12. A. Loveday (1914). The History and Economics of Indian Famines.

13. I.F.C.R. of 1880, p. 35.

14. S. Brahme (1983). *Drought in Maharashtra*, p. 107.

15. S. Liberrnan (1984). Field Level Perspectives on the administration of Maharashtra E.G.S.

16. Walkas & R. P. Singh (1986). *Risk benefits crop insurance and dryland agriculture*, E.P.W. June 1986.

17. Hanumantrao (1988). *Unstable Agriculture and Droughts*, p. 92.

18. I. Hirway (1985). Wage Employment Programme in Rural Development Given in Hanumantrao, p. 96.

19. G.O.I., *Ministry of Programme Implications*, June 1987.

20. I. Hirway, *op. cit.*

21. Business Standard, *Drought Prone Area Programme in India* by Mund in February 1985.

22. *Ibid.*

23. G.O.I. *Seventh Five Year Plan*, Vol. II, p. 61.

24. Business Standard, *op.cit.*

25. Samant and Mishra (1978). *Evaluation of DPAP*, Baroda, Given in Hanumantrao, p. 87.

26. S. Brahme(1983). *Drought in Maharashtra 1972*, p. 122.

27. S. Brahme (1990). Land Use in Western Zone of India: Problems and Perspective, Chaps. V and XI.

5

Assessment and Impact of Droughts 1987-2002

It is gradually being recognised that the current drought is probably the century's worst. It certainly is the most serious problem confronting the country today. Inadequacy of the monsoon is a matter of great concern this year and a large part our country is facing security drought situation. In some States, this is the third or fourth successive drought year. About a third, if not more, of India's rice, pulses and oilseeds growing areas, and over one-half of the lands growing coarse grains—which together with the former account for the overwhelming proportion of the kharif crop—are now in drought's dealy grip. Sowing has been delayed in half the cultivated area. Only four of the country's 47 major reservoirs have water level that are 70 per cent as high aslast year's.

The Agricultural Drought

The failure of monsoon this year is more than what experts describe as 'meteorological drought', *i.e.*, a decrease of 25 per cent or more in the normal percipation. Indeed, it is worse than even 'hydrological drought, *i.e.*, prolonged meteorological drought, with a marked depletion of surface water, drying up reservoirs, lakes and streams, cessation of spring flows and fall in ground water levels.

Going by the definition given by the National Commission on Water, most of the country is now witnessing a full-fledged 'agricultural drought less than half the normal weekly rainfall for four consecutive weeks in the period mind-May to mid-October. In an agricultural drought, soil moisture and rainfall are too scarce to be able to support the healthy growth of crops during the critical phase of their growth cycle-Extreme crop stress and wilt result.

Consequences

They likely consequence of such a drought are truly horrifying-immediate shortage of foodgrains (barring wheat and rice from the bufferstock) vegetables, fodder and milk; cattle deaths; ruination of millions of small peasants; over-exploitation and further devastation of land; double-digit inflation and growing political unrest.

Due to drought conditions prevailing in the country, kharif crops in most of the States have suffered and its output has been considerably reduced. In certain parts of the country the farmers have lost most of them production. This has resulted in lower yields as well as very little or no work to the agricultural labourers. All this has resulted in reduced incomes of the farmers and others dependent on farming as their livelihood. This as resulted in a state of some kind of economic depression in the country.

Table 5.1

Rainfall Pattern in Different Meteorological Divisions

Years	*1987*	*1986*	*1985*	*1984*	*1983*	*1982*
Week ending on	*9/9*	*10/9*	*11/9*	*12/9*	*7/9*	*8/9*
1. Excess	2	X	2	4	10	X
2. Normal	11	19	22	21	21	20
3. Deficient	19	16	11	10	4	14
4. Scanty	3	X	X	X	X	X
5. No Rain	X	X	X	X	X	X
6. Data Inadequate	X	X	X	X	X	X
Total	35	35	35	35	35	35

Table 5.1 gives the rainfall pattern and the number of meteorological sub-divisions falling into each category. It is to be seen that during the year 1987 only 2 sub-divisions received excess rainfal, 11 normal and 19 deficient. Three sub-divisions reported scanty rainfall. Whereas in earlier years, no sub-division has reported deficient rainfall was also less.

The Rainfall

The cumulative rainfall data shows that till September 16, 1987 only 12 out of the total 35 meteorological sub-divisions had received normal or excess rains. Last year, in the corresponding period, 19 sub-divisions had received normal or excess rainfall.

It would be seen from Table 5.2 that out of 384 districts for which rainfall data are available for the period 1-6-1987 to 9-9 1987 only 31 districts have reported excess rainfall while 122 districts have reported normal rainfall. The rainfall was deficient in 158 districts and scanty in 52 districts. It is also to be noted that during this year only 43 per cent of all the districts have reported excess or normal rainfall as compared to 46 per cent of the falling in this category in the drought year of 1982, Fourteen per cent of the districts this year has reported scanty, while 43 per cent of the districts have reported deficient rainfall:

Table 5.2

No. of Districts with Excess, Normal, Deficient and Scanty Rainfall for the period 1st June, 1987 to 9th September, 1987

State	*Excess*	*formal*	*Deficient*	*Scanty*	*N.R.*	*	*Total*
Andhra Pradesh	2	14	6	1	–	–	23
Assam	3	7	5	1	–	5	21
Bihar	12	18	2	–	–	1	35
Gujarat	–	3	8	8	–	1	20

State	*Excess*	*formal*	*Deficient*	*Scanty*	*N.R.*	*	*Total*
Haryana	—	—	5	8	—	—	13
Himachal Pradesh	—	1	7	3	—	1	12
Jammu & Kashmir	1	2	1	—	—	6	10
Karnataka	3	11	5	—	—	—	19
Kerala	—	4	9	1	—	—	14
Madhya Pradesh	3	16	26	—	—	—	45
Maharashtra	—	19	8	1	—	—	28
Orissa	—	3	10	—	—	—	13
Punjab	—	1	5	6	—	—	12
Rajasthan	—	2	12	12	—	1	27
Tamil Nadu	—	8	10	1	—	1	20
Uttar Pradesh	1	3	39	10	—	3	56
West Bengal	6	10	—	—	—		16

Legend : Excess : +20% or more
Gormal : +19% to- 19%
Deficient : -20% to - 59%
Scanty : -60%to - 99%
No Rain : -100%
Data inadequate

Total number of districts out of 384 from which data have been received = 363. Total No. of station from which data have been received = 1384.

District-wise percentage Distribution

	1982	*1983*	*1984*	*1985*	*1986*	*1987*
Excess	9	31	16	15	9	9
	46	75	63	62	50	45
Normal	37	44	47	47	41	34
Deficient	50	24	35	35	47	43
Scanty	4 54	1 25	2 37	3 38	3 50	14 57
No Rainfall	0	0	0	0	0	0

Other Drawbacks

Reports indicate that the kharif sowing operations have been seriously hampered in many parts, resulting in a substantial shrinkage in the net sown area. The standing crops in many pockets are showing signs of roistered stress. Scarcity of drinking water and fodder has been reported from several States. The milk yield of cattle has declined. Due to water scarcity, the cattle have to drink dirty water from the village ponds, which has resulted in serious cattle diseases and lot of cattle have died. The output of almost all the kharif crops is likely to fall. The worst hit States are: Haryana, Punjab, Rajasthan and Gujarat. The zones severely affected include other notably productive areas like West U.P., coastal Andhra Pradesh, Central Bihar, Saurashtra and Kutch.

Situation in the States

In Haryana paddy transplantation has taken place in only 40 per cent of the rice cultivable areas. Monindergarn and Bhrwani districts are facing complete drought and have serivce drinking water problem. The Bajra crop has been completely ruined. In Punjab, paddy has not been transplanted over a third of the normal area. In this State, which has a high per centage of irrigated area, farmers are not currently getting more than 8 hours of power. In Gujarat, there has been no sowing in 78 of the 182 talukas in the State. Crops like paddy, cotton, purses, bajra and maize are considered totally lost with little or no chance of recovery. The State Government fears that the kharif output could be the lowest in 25 years. The situation in Rajasthan, which is faced with the fourth-successive drought is just as bad. As many as 26 out of 27 districts have had practically no rainfall. Kharif sowing was done in barley a third of the target area. In U.P., as many as 50 out of 57 districts area in the grip of a severe drought. Half of the croped area of 189 lakh heactares remains unsown. Bajra, paddy and maize are the word hit crops.

The situation is hardly better in Madhya Pradesh. Of the State's 45 districts, a third have received between 10 and 30 per cent of the normal rainfall while 16 others have much less than half the average. In Maharashtra, the only redeeming feature is that the seven districts of Marathwada, affected with a drought for last three years, have had good and uniform rain. But six other traditionally drought prone districts in central Maharashtra and almost all of Vidarbha have been starved of rain after an initial wet spell. The overall situation is much the same in the four southern States, Karnataka yet to recover from four successive years of drought seems to be heading for another in its southern parts. Kharif prospects in Andhra Pradesh the largest producer of paddy in the country are distinctly bleak, June and July are the peak sowing season but with rainfall deficient or scanty in all but one (Guddapah) of the State's 18 districts, the average has fallen even in the rice-bowl districts of East and West of Godawari. Of Kerala's 14 districts, only 4— Alleppey, Ernakularn, Kottayam and Trichur—have received rainfall. The Kharif paddy, banana and other crops may be affected in the southern most districts. In Tamil Nadu, the north-east monsoon is the main source of rainfall which is less than satisfactory. Normally, the Mettur Dam is open in early June but because of the poor inflow, it could well be delayed as last year.

The eastern states, in comparison have had a good monsoon. Assam, Meghalaya, West Bengal, Sikkim, Orissa and the Bihar Plains have had excess or normal rainfall, Arunachal Pradesh, Nagaland, Manipur, Mizoram and Tripura had deficient or scanty rain but since then average is normally far in excess of the amount of rain required for the kharif crops, the deficiency may not affect the output very much.

The (efforts) in India to tackle the problem of made to assess the crop production prospects for different crops:

Rice

Eastern States account for 69 per cent of the kharif rice area. There is acreage shrinkage due to drought in the States of Uttar Pradesh, Madhya Pradesh and Orissa and due to prolonged floods in Assam, Bihar and West Bengal. 'Biasi' or 'Bushning' operation could not be done in the upland rice areas in Madhya Pradesh and Orissa as a result fields are heavily infested with weeds and the crop growth is stunted. Crop yields in such a situation will be reduced. In the flood affected crops; the plant population is thin and top dressings could not be undertaken properly. This will also lead to reduced production.

The Southern regions of Andhra Pradesh, Karnataka, Tamil Nadu, Maharashtra and Kerala together account for about 20 per cent of the rice area in the country. Due to abberent monsoon, there is area shrinkage in the region too. The crops have also suffered intermittent moisture stress during the growth period.

The nortern region comprising of Punjab, Haryana, Himachal Pradesh, Jammu & Kashmir accounts for about 10 per cent of the rice area in the country. In spite of the fact that Punjab and Haryana grow rice under irrigation, there has been heavy shrinkage in area in both these States. The water table has also gone down. Due to insufficient rains, the rice crop in Jammu and Kashmir and Himachal Pradesh is also unsatisfactory.

As against the kharif rice output of 56.20 million tonnes in triennium ending 85-86 kharif seasons, we may have an output of around 47.45 million tonnes in the kharif 1987.

Coarse Grains

The average area under coarse grains is about 33 million hectares. The important crops are Jowar, Bajra and Maize. The important State where coarse grains are cultivated are Maharashtra, Rajasthan, Karnataka, Uttar Pradesh, Madhya Pradesh, Andhra Pradesh, Gujarat, Tamil Nadu, Haryana and Dinar. The overall pattern this year has not been favourable for these crops. The sowing was done late and hardly 60 per cent of the area is reported to have been covered.

Jowar

In Maharashtra, which grows about 30 per cent of the country's jowar during kharif season, the rainfall from the beginning of Sout West monsoon was favourable for jowar cultivation in the State. The rainfall was also good for jowar sowings in Karnataka but it remained erract during July, 1987 which created moisture stress. In Andhra Pradesh, the position improved from first week of July. The rains received after the second week of July in the States of Bihar, Uttar Pradesh, Madhya Pradesh and Gujarat were also helpful: for the farmer for taking up further sowings. Thus, the timely sown crop suffered due to moisture stress in the month of July upto middle of August. The late sown crop would also yield less.

Bajra

In the main bajra growing states, the rainfall is not favourable. Rajasthan is the largest Bajra growing State accounting for about 4.7 million hectares, out of the total 10.6 million hectare area in the country. Haryana and Gujarat are the other important Bajra growing States. Due to delay and weak monsoon large scale sowings of Bajra were not possible. Even the sown crop was affected due to acute water stress. In some of the areas the crop withered away. However, rains in the month of August helped in retrieving the crop to some extent. Mainly it has helped in raising bajra as a fodder crop in areas where the main bajra crop had failed completely. The overall crop situation for bajara is not satisfactory in the country.

Maize

Maize is an important crop in the State of Uttar Pradesh, Rajasthan, Madhya Pradesh and Bihar. In all the States the monsoon was late. The rains received from the first week of July in U.P. and Bihar were beneficial to maize crop. Similarly, the rains received from 2nd week of July in Himachal Pradesh were helpful. The delay in monsoon affected timely sowings of the crop in other States. In Jammu and Kashmir and Himachal Pradesh, normally the sowings are over by the end of June but till the end of July, the rainfall was not favourable which affected the sown crop adversely. The good rains in August saved the crop from withering away. The recent floods in Bihar have also damaged the standing maize crop.

It may be seen that the overall prospects of coarse grains production in the country during the current Kharif season are not satisfactory.

It is reported that the kharif coarse grains production was 25.51 million tones for the triennium ending 1985-86 which may be around 19 million tonnes during the kharif of 1987.

Pulses

Coverage under kharif pulses has been reported at 15.16 lakhs in Madhya Pradesh, 10.44 lakh ha in Maharashtra, 6.48 lakh ha. in Karnataka and 8.01 lakh ha. in Andhra Pradesh and 5.36 lakh ha. in Gujarat. The coverage under pulses in Rajasthan is reported at 3.85 lakh ha., Orissa 4.27 ha., Bihar 3.59 lakh ha. and U.P. 4.62 lakh ha. The crop had a initial setback due to drought conditions in Rajasthan, Gujarat, Uttar Pradesh and part of Orissa. The crop situation in other States in more or less normal. In the States which had the drought conditions there has been crop mortality ranging from 50 to 100 per cent. The late sowing of short duration pulses like *Moong* and *Urd* are still being done in flood affected. States of Bihar, West Bengal and Assam. The sowing of these crop has also been taken up in Orissa. Basically there is very little loss to early maturing *Arhar* crop which has a high degree of drought resistance. The rains in the month of September have considerably improved the conditions of the standing crops in the States of Madhya Pradesh, Maharashtra, U.P., Rajasthan, Gujarat etc. It is anticipated that the exent of shortfall in the production of pulses may be hardly anything less than that obtained in triennium ending 85-86 (4.8 million tonne).

Oil Seeds

Sowing of groundnut have badly been affected due to erratic monsoom behaviour in the country. In Gujarat, 4 lakh hectares, out of 19 lakh hectares have been sown and 40 per cent of the sown crop has been damaged. Similarly, in Andhra Pradesh, 3 lakh hectares out of 14 lakh hectares have been sown. In Maharashtra, only 6.1 lakh hectarse have been sown. In Karnataka, where the area under groundnut is 8 lakh hectares, only 30 per cent has been covered. In Orissa 80 per cent area out of 1.7 lakh hectare have been covered under groundnut. In U.P., 60 per cent groundnut has been damaged due to drought. The groundnut will be about 15 to 20 per cent loss during the kharif due to crop damages in many states. In Madhya Pradesh soyabeen has been sown in 11 lakh hectare.

Cotton

Drought has seriously affected sowing of cotton and vegetative growth in Gujarat State. In the irrigated areas of Punjab, Haryana and Rajasthan also crops has been slightly affected due to non-availablility of required number of irrigations. The crop condition in Southern states is normal. Sowing in Tamil Nadu are yet to be undertaken. No severe attack of pests and diseases has been reported from any States. As the day time temperature is quite high, the ball opening is taking place and thus there is likely to be a reduction in the yield of the crop. The crop size, however, would be around last year level.

Jute

It is estimated that an area of about 6.52 lakh ha. under jute and 2.0 lakh ha. under mesta has been shown in different jute mesta growing States. The jute area of 21 thousand ha. has been estimated to be fully damaged while 33 thousand ha. of jute area have been estimated to be partially damaged by flood waters in Assam. In Uttar Pradesh 12.4 thousand ha. of jute area is reported to bedamaged by drought. Harvesting of jute in most of the States is in full swing.

Sugarcane

The area covered under sugarcan during 1987-88 is estimated to be 21.22 lakh ha. Drought has affected sugarcan crop in Rajasthan, Punjab, Haryana, U.P. and Madhya Pradesh. The recovery per centage in these areas may be higher but the yield level is goving to be lower as compared to last year.

Thus, the total kharif output during 1987 would be around 70.2 million tonnes, comprising of 47.45 million tonnes of rice, 19.07 million tonnes of course cereals and 3.68 million tonnes of pulses. In the case of kharif oilseeds, the expected output would be around 5 million tonnes.

Alternative Crop Strategy

In order to mitigate the affect of aberraions in the South-West monsoon 1987-88, certain strategies have been developed which are to be followed in different regions depending upon he severity of the drought conditions.

As the monsoon progresses or as the time passes on in the kharif season, the growing season would be reduced. In other words the same crop or variety of crops may not be suitable for later periods than the nomal. This means alternate crops varieties have to be choosen to be sown in later dates. Clusters bean Cowpea and Greengram can be taken up. Another important consideration is the need to increase the seed rate by 15.25 per cent in order to compensate the reduced growth of the late sown crops.

Mild Season Corrections

In the Peninsular India, Eastern parts of the country and parts of North-West region, crops have already been sown. But due to the skewed distribution of rainfall they might have been under moisturous stress. Under these circumstances, in order to increase the productivity of the standing crops some of the suggestions are:

(a) If an extended stress occure immediately after sowing very little can be done to save the crop.

(b) On the other hand, if mild stress occurs, it can be overcome by mulching the soil with the help of Blade Harrows or country Ploughs.

(c) If the stress period extends, the crop naturally would wilt and suffer actuely. In such a case, crops like groundnut and arhar can be saved partly by spraying having 2-4 per cent dilute urea solution in the evening house.

Compensatory Programmes

When Kharif 1987 is not too adequate, emphasis should be more on the Rabi season where in increased area is under irrigation. Hopefully with the subsequent good rains, it might be possible to have adequate stored soil moisture for raising good crops even under rained conditions. However, to optimize irrigation for other crops, re-scheduling may be done so that the most sensitive growth stages of the crop are adequately covered with sufficient supply of irrigated water.

The watering of wheat crop has to be suitably planned based on the water availability in the area. If the water is only available for two irrigations, then the second irrigation should be given after 100 days of planting at the flowering stage of the crop growth. If the availability of water is for three irrigation, second irrigation may be given after 75 days of planting at the jointing stage of the crop. In order to compensate the loss in kharif output, ail out efforts are being made to achieve the target of 70-72 million tonnes foodgrains production during rabi season of 1987-88.

Seed Programme

In a season of the present type, one important casualty could be the seed for the next year. Efforts necessarily have to made to organize heavy seed production programmes under protective irrigation conditions under various major and minor irrigation systems. The Department of Agriculture at the Centre, in collaboration with the State Governments has drawn contingency plans for the various affected States and is assisting them in procurement of seeds as well as monitoring of inter-state movement of seeds.

Folder Banks

Under aberrate situations with long break of monsoon, fodder for animal also could become a serious bottleneck. It would be necessary to look for alternate source of fodder under delayed kharif sowings. Cereals like Sorghum (MP Chari) and short pulses. Like Horse gram (Kulti) should be shown. They would provide the needed fodder extent. In other words, no land should be left fallow just because the conventional crops of the retion could not be sown due to the delayed receipt rain. Department of Agriculture & Cooperation has sent Teams of is animal husbandry exparts with Department of Forests representatives to the works affected States to assess their requirement of fodder for the next months and to locate source of fodder availability.

What the Centre is Doing

At the Centre, the Cabinet Committee on Drought has been set up under the Chairmanship of the Prime Minister. The Prime Minister has written to the Chief Ministers/ Governors of the States/Union Territories regarding the steps to be taken on a number of fronts including crop production, provision of employment, fodder, power, irrigation, drinking water and supply management for essential commodities to fight the ourrent drought. The Cabinet Committee on drought sanctioned acquisition of 17 special rigs by Central Ground and Water Board at a cost of Rs. 15 crores. The rigs will be procured within four months and deployment in States facing acute drinking water problem.

A special scheme has been sanctioned for growing short duration fodder crops by small and marginal farmers in 2.25 lakh hectares to produce an expected 60 lakh tonnes of given fooder.

The scheme, which will cost Rs. 9 crores, is to be implemented in Gujarat, Madhya Pradesh, Rajasthan, Tamil

Nadu, Bihar, Haryana and Punjab and provides for subsidies to be given at the rate of Rs. 500/ha. for small farmers and Rs. 400/ha. for marginal farmers for inputs. The States are expected to provide irrigation free or at a confessional rate. Fodder crops will be grown without interfering with rabi cultivation.

It is expected that the scheme will improve fodder availability in the affected areas and will obviate the need to transport fodder over long distances.

The Government had deferred recovery of all bank loans from farmers in drought affected areas. All commercial banks have been asked to defer recovery for a period of two years or till the next normal year, whichever, is earlier.

Banks have also been directed by the Reserve Bank not to charge penal interest in respect of amounts faling due in the current year, nor to compound interest on the dues postponed in pursuance of these AEI instuctions.

The failure of the monsoon during the current year has brought a new aucertainty to economic outlook, which so far had seemed to indicate a recovery. In order to get some idea about the extent of possible damages in a bad agri-growth year, we looked at the historical data for past 16 years. The data suggest the poor agri-growth year have typically had a larger impact on the growth rates than on inflation. During 1986-87 to 2001-2002, there was a drop in agricultural GDP for six years.

However, the impact was more on the foodgrains outputs in these years—particularly on kharif crop. During these six years, the kharif foodgrains output dropped on a average by 5.2 per cent. The average drop in foodgrains output at 4.5 per cent was somewhat moderated by the rabi foodgrains, whose output dropped by 3.3 per cent on an average. But the average drop in agricultural GDP in these years—at 1.2 per cent—was much lesser than in foodgrains, reflecting the diversification into now grain crops. According to report, 1997-98 was the worst year for the sector, when agricultural GDP had declined by 2.4 per cent.

Going by the past experience, if one expects a drop of

1.2 per cent in agri-GDP in 2002-03, then it would lead to a direct drop of 0.3 to 0.5 per cent in overall GDP. In the last fiscal, agriculture had contributed 1.4 per cent to GDP growth. A reversal from that position would mean, in an incremental sense, a drop of 1.5 to 2 percentage points in the economy's groywth rate.

Apart from this direct impact, there could be some indirect impact on growth too, to the extent that the rest of the economy depends on rural demand. In the past, the performance of the non-agricultural GDP has not been distinctly different in the good and bad agricultural years.

DROUGHT-2002

The major droughts year in India were 1877, 1899, 1918, 1972, 1987 and 2002. Large parts of the country perennially reel under recurring drought, over 68 per cent of India are vulnerable to drought. The 1987, drought was one of the worst droughts of the country with an overall rainfall deficiency of 19 per cent.

It is gradually being recognized that the current drought is probably the century's worst. It affected 59 60 per cent of the crop area and a population of 285 millions. It certainly is the most serious problem confronting the country today. Inadequacy of the monsoon is a matter of great concern this year and a large part our country is facing severe drought situation.

The failure of monsoon this year is more than what experts describe as meteorological drought i.e., a decrease of 25 per cent or more in the normal precipitation. Indeed, it is worse than even hydrological droughts, i.e. prolonged meteorological drought, with a market depletion of surface water, drying up reservoirs, lakes and streams, cessation of spring flows and fall in ground water levels. Going by the definition given by the National Commission on water, most of the country is new witnessing a full-fledged agricultural drought less than half the normal weekly rainfall for four consecutive weeks in the period mid-May to mid-October. In an agricultural drought soil moisture and rainfall are too scare to be able to support the healthy growth of crops during the critical phase of their growth cycle extreme crop stress and wilt result.

The likely consequence of such a drought are truly horrifying immediate shortage of food grains vegetables fodder and milk, cattle deaths, ruination of millions of small peasants, over population and further deviation of land, double digit inflation and growing political unrest. Due to drought conditions prevailing in the country Kharrif crops in most of the states have suffered and its output has been consideration reduced. In certain parts of the country, the farmers have lost most of them productions. This has resulted in lower yields as well as very little or no work to the agricultural labourers. All this has resulted in reduced incomes of the farmers and other dependent on farming as their livelihood. This has resulted in a state of some kind of economic depression in the country.

Table 5.3 gives the rainfall pattern and the number of meteorological subdivisions falling into each category. It is to be seen that during the year 1987 only 2 sub-divisions received excess rainfall, 11 normal and 19 deficient. Three sub-divisions reported scanty rainfall. Whereas in earlier year, no subdivision has reported deficient rainfall was also less.[1]

Table 5.3
Rainfall Pattern in Different Meteorological Divisions

Years	*1987*	*1986*	*1985*	*1984*	*1983*	*1982*
week ending on	*9/9*	*10/9*	*11/9*	*12/9*	*7/9*	*8/9*
1. Excess	2	x	2	4	10	X
2. Normal	11	19	22	21	21	20
3. Deficient	19	16	11	10	4	14
4. Scanty	3	X	X	X	X	X
5. No Rain	X	X	X	X	X	X
6. Data Inadequate	X	X	X	X	X	X
Total	35	35	35	35	35	35

The Rainfall

The commulative rainfall data shows that till September 16, 1987 only 12 out of the total 35 meteorological sub-divisions had received normal or excess rains. Last year, in the corresponding period, 19 subdivisions had received normal or excess rainfall.

It would be seen from Table 5.4 that out of 384 districts for which rainfall data are available for the period 01.06.1987 to 09.09.1987 only 31 districts have reported excess rainfall while 122 districts have reported normal rainfall. The rainfall was deficient in 158 districts and scanty in 52 districts. It is also to be noted that during this year only 43 per cent of all the districts have reported excess or normal rainfall as compared to 46 per cent of the falling in this category in the drought year of 1982. 14 per cent of the districts this year has reported scanty, while 43 per cent of the districts have reported deficient rainfall:

Table 5.4
No. of Districts with Excess, Normal, Deficient and Scanty Rainfall for the Period 1st June, 1987 to 9th September, 1987

State	*Excess*	*Normal*	*Deficient*	*Scanty*	*N.R.*	*	*Total*
Andhra Pradesh	2	14	6	1	–	–	23
Assam	3	7	5	1	–	5	21
Bihar	12	18	2	–	–	3	35
Gujarat	–	3	8	8	–	1	20
Haryana	–	–	5	8	–	–	13
Himachal Pradesh	–	1	6	3	–	1	12
Jammu and Kashmir	1	2	1	–	–	6	10
Karnataka	3	11	5	–	–	–	19
Kerala	–	4	9	1	–	–	14

State	*Excess*	*Normal*	*Deficient*	*Scanty*	*N.R.*	*	*Total*
Madhya Pradesh	3	16	26	—	—	—	45
Maharashtra	—	19	8	1	—	—	28
Orissa	—	3	10	—	—	—	13
Punjab	—	1	5	6	—	—	12
Rajasthan	—	2	12	12	—	1	27
Tamil Nadu	—	8	10	1	—	1	20
Uttar Pradesh	1	3	39	10	—	3	56
West Bengal	6	10	—	—	—	—	16

Source: India Meteorological Department Pune.

Table indicates that the *kharif* sowing operations have been seriously hampered in many parts, resulting in a substantial shrinkage in the net shown area. The standing crops in many pockets are showing signs of moisture stress. Scarcity of drinking water and fodder has been reported from several States. The mild yield of cattle has declined. Due to water scarcity, the cattle have to drink dirty water from the village ponds, which has resulted in serious cattle diseases and lots of cattle have died. The output of almost all the *kharif* crops is likely to fall. The worst hit States are Haryana, Punjab, Rajasthan and Gujarat. The zones severely affected include other notably productive areas like West U.P. coastal Andhra Pradesh, Central Bihar, Saurashtra and Kutch.

The most recent, the drought of 2002, ranks fifth in terms of magnitude but is unique when examined in overall terms of magnitude, spacing, dispersion and duration. In July 2002, rainfall deficiency dropped to 51% surpassing all previous droughts. The impact of drought spread over 56% of the landmass and threatened the livelihoods of 300 million people across 18 states. The 2002 monsoon was one of the shortest in recorded history.

Monsoon Performance: Indian meteorological department (IMD) defines droughts as events when rainfall during full south west monsoon season (June 1 to September 30) for the country as a whole, is deficient by more than 10 per cent of the long term average and when more than 20 per cent of the country's area is affected by drought conditions. In 2002, for the country as a whole the full season rainfall has been 19 per cent below normal and 29 per cent of all India area has been affected by drought.

All meteorological subdivisions and districts levels the IMD adopts the following criteria for characterizing monsoon—

Excess	:	+ 20% of LTA or more
Normal	:	+ 19% of LTA
Deficient	:	- 20% to - 59 % of LTA,
Scanty	:	- 60% of LTA or less.

At the end of year 2002, monsoon season 21 subdivisions had received deficient / scanty rainfall, same as that in 1987. In 2002, two of these meteorological subdivisions (East and West Rajasthan) scanty rainfall compared to three in 1987. However, for assessing the drought conditions at meteorological subdivision district level 31, IMD defines a moderate meteorological drought which rainfall is 26-50 per cent below LTA and severe meteorological definition, 12 subdivisions are drought affected. They are listed in Table 1.1 below. Total area of these 12 sub-works out to be 29 per cent of which 10 per cent fall under the category of severe drought and 19 per cent under moderate drought. Nine other subdivisions received deficient rainfall in the range of 20 to 25 per cent of the LTA. These are Himachal Pradesh, West U.P. East U.P., A.P., West M.P. Gujarat, Daman & Diu, Saurashtra, Kutch & Div, Telengana and Andman & Nicobar Islands.

The drought 2002 is indeed widespread and the table above masks some important deleterious features of the monsoon failure this year. *First,* parts of the country which fall outside the 21 subdivisions that are characterized as

Table 5.5
Sub-Divisions Affected by Meteorological Drought

Meteorological Sub-Division	*Full-season rainfall as % of LTA*
Severe Drought	
1. West Rajasthan	–71
2. East Rajasthan	–60
Moderate Drought	
3. Tamil Nadu & Pondicherry	–45
4. Lakshadweep	–45
5. South Interior Karnataka–44	
6. Haryana, Chandigarh & Delhi	–36
7. Punjab	–36
8. Kerala	–35
9. Rayalaseema	–33
10. North Interior Karnataka	–31
11. Coastal Karnataka	–30
12. Coastal Andhra Pradesh	–26

Source: India Meteorological Department.

deficient/scanty have also been hit by poor rainfall. For example, North Eastern States of Meghalaya, Nagaland and Mizoram have received cumulative rainfall that is less by 30 per cent or more of the LTA, but they fall under sub-divisions that have received normal rainfall. Same is the case with State of Goa. At district level, several districts falling under normal sub-divisions have received deficient rainfall. *Second,* the characterization of States/Regions as drought affected on the basis of full-season rainfall masks the temporal aspects of the rainfall. Several parts received good monsoon in August and also September, but were hit almost irretrievably by severe dry spell in July. For the country as a whole, rainfall deficiency in July turned out to be a whopping 49 per cent; the worst recorded for this month at least in last 100 years. Only on two occasions in the past had the July deficiency exceeded 45 per cent—in 1911 and in 1918. For Central India the deficiency in July exceeded 75 per cent. *Third,* the full season rainfall data that defines meteorological drought as per IMD's criteria does not fully captures the July impact.[2] The hydrological drought may also have been lessened by good rains in August and to some extent in September. However, the drought 2002 in terms of agricultural drought has been rather severe Inadequate precipitation and soil moisture stress coupled with July monsoon failure has affected the prospects of area, production and yield for agricultural crops severely this year, So, while in meteorological terms drought 2002 has been less intense than drought 1987, the same cannot be said for agricultural drought. If anything, the agricultural drought 2002 can be expected to be more severe for *Kharif* crops than the drought 1987, because July is critical for agricultural operations in most parts of the country.

The revival of monsoon in the first week of August was pronounced and sustained through the remaining weeks of the month. Much of the deficiency from the central parts of the country was made good by August rainfall. Orissa, Chhattisgrah, Interior Maharashtra, Gujarat and Telangana received plenty of rainfall towards the end of August and first few days in September. September rains were not as good as august, but were still a factor in reducing the cumulative deficiency caused by July dry spell. The withdrawal of monsoon in September was delayed somewhat. Mid-September brought good rains to central and northern parts of the country. Rainfall was also heavy during last week of September in Eastern and northeastern part of the country.

Monsoon Performance For Full Monsoon Season (June 1 to end-September)

For the full South-west monsoon season in 1986 (June to September), 14 sub-divisions received deficient rainfall. While no sub-division received scanty rainfall, while 21 sub-divisions received normal rainfall. Similarly, by the end-September 1987. 14 sub-divisions received deficient rainfall and 3 subdivisions received scanty rainfall, while 12 sub-divisions received normal and 2 sub-divisions received excess rainfall. During the South-west monsoon season 2002, as many as 19 sub-divisions received deficient rainfall and 2 sub-division received scanty rainfall, while 14 subdivisions received normal rainfall and 1 sub-division (Jammu & Kashmir) received excess rainfall.[3]

The inter-temporal progress of monsoon week-by-week is given in Table 5.6.

Flood Situation

Analysis for flood situation during 2002 indicates that while parts of Assam and Bihar were reeling under floods for several weeks, floods hit but have receded in Arunachal Pradesh, Gujarat, Himachal Pradesh, Kerala, M.P., Maharashtra, U.P. and west Bengal. In Bihar, some parts of state were badly flooded due to heavy rains in Nepal, while some others experienced drought as a result of inadequate rainfall in the area of Sate. Ministry of Home Affairs, as the nodal ministry, is coordinating the central response to this calamity like situation. The rescue and relief measures assumed the form of evacuation of persons from flood-affected parts, medical aid, supply of rations, and distribution of food packets in Assam and Bihar. State-wise details of damage due to cyclone, heavy rains and floods during South-West monsoon 2002 is provided in Table 5.7

Agricultural Position

The Prospects for Indian agriculture were adversely affected by deficient monsoon after 13 continuous years of

Table 5.6

Inter-temporal Progress of South-West Monsoon

Cumulalive Rainfall upto	*Excess*				*Normal*				*Deficient*				*Scanty*			
	1986	*1987*	*2001*	*2002*	*1986*	*1987*	*2001*	*2002*	*1986*	*1987*	*2001*	*2002*	*1986*	*1987*	*2001*	*2002*
June 1 to July 31																
2nd week of June	6	7	25	11	1	6	8	7	8	16	2	11	20	6	0	7
3rd week of June	8	11	25	8	3	9	5	19	14	13	5	7	9	2	0	2
4th week of June	13	6	20	14	6	7	9	13	10	20	6	8	5	2	0	1
1st week of July	9	3	17	10	12	14	12	15	10	14	6	11	1	4	0	0
2nd week of July	7	1	16	6	14	15	12	10	11	13	6	20	0	6	1	0
3rd week of July	4	2	16	0	16	10	11	14	14	20	8	18	1	3	0	4
4th week of July	5	0	13	2	15	10	10	14	14	20	11	16	0	5	1	4
5th week of July	7	0	10	1	10	10	16	9	17	21	9	21	0	4	0	5
August 1 to September -30																
1st week of July	5	2	10	1	15	7	16	10	14	21	9	21		5	0	0
2nd week of July	4	3	6	1	17	7	22	16	12	21	7	18		4	0	1
3rd week of July	6	3	5	1	16	9	23	15	12	20	7	17		3	0	3

Cumulative Rainfall up to	*Excess*				*Normal*				*Deficient*				*Scanty*			
	1986	1987	2001	2002	1986	1987	2001	2002	1986	1987	2001	2002	1986	1987	2001	2002
4th week of July	2	2	4	1	21	11	22	16	12	18	9	16	0	4	0	3
1st week of July	0	2	1	0	19	11	26	19	16	19	8	16	0	3	0	1
2nd week of July	0	2	1	0	19	10	25	20	16	20	9	15	0	3	0	1
3rd week of July	0	2	1	1	19	12	27	19	16	18	7	15	0	3	0	1
4th week of July	0	2	1	1	21	12	29	14	14	18	5	20	0	3	0	1
June 1 to September 30	0	2	1	1	21	12	29	14	14	18	51	19	0	3	0	2

Source: India meteorological Department, Pune.

Table 5.7
Damage due to Cyclone, Heavy Rains and Floods During South-West Monsoon 2002

State/Union Territory	Affected				Damage	
	District	Villages	Total (in Lakh Hectares	Population (in Laks)	Crops Area (in Lakh Heactares	Estimated Valuse of Crops (in Rs. Crore)
Arunachal Pradesh–FFL	3 (13)	10	NR	0.1580	Neg.	Neg.
Assam – HR/L	21 (23)	6226	8.036	53.86	3.23	NR
Bihar – HR	22 (38)	7314	17.18	140.34	10.43	113.24
Gujarat – HR	10 (25)	129	NR	6.50	NR	NR
Himachal Pradesh – HR/L	1 (12)	NR	NR	NR	NR	NR
Kerala-HR/F/L	7(14)	680	NR	0.21	NR	1.42
Madhya Pradesh – HR	1(45)	2	NR	NR	NR	NR
Maharashtra – HR/L	1(35)	311	NR	NR	NR	NR
Uttar Pradesh – HR	2 (70)	42	0.14	0.75	0.11	0.10
West Bengal – HR/L	4	115	NR	0.26	Neg.	0.21

F = Flood   FF = Flash Flood   L = Land Slide
HR= Heavy Rains   NR = Not Reported
Figures in parenthesis indicate total number of districts.

normal rainfall, reflecting the continued rain depending of Indian agriculture. The country as a whole received about 81 per cent of the long period average (LPA) rain fall during the south west monsoon season in 2002-05 compared with 90 per cent in 2001. As a result 21 meteorological subdivision out of 36 received either deficient or scanty rainfall in comparison with 5 sub-divisions in the previous monsoon. According to the India Meteorological Department (IMD), the shortfall in rainfall during the 2002 south west monsoon season led to drought in as much as 29 per cent of the land with 19 per cent being declared as moderate drought and 10 per cent as severe drought areas. Further more, the scale and intensity of the drought in 2002 has been comparable with that in 1987-88 when 21 sub-divisions received deficient/ scanty rainfall and 14 sub-divisions received excess/normal rainfall.

As a result of drought conditions, the real GAP agriculture and allied activities is estimated to have declined by 3.1 per cent in 2002-03 in contrast to a rise of 5.7 per cent in 2001-2002[4].

The overall performance of agriculture production has been affected adversely by the severe drought in 2002. Production of foodgrains in 2002-03 is estimated to decline sharply to 183.2 million tonnes from the record production of 212.0 million tonnes in the previous year.

The wholesale price index of foodgrains, in general, and that of cereals, in particular, remained stable during the post south west monsoon season, which was affected by drought. The target for the foodgrain production for the year 2002-03 has been set at 220 million tonnes but achievement is only 183.17 million tonnes. It means that the total foodgrains during 2002-03 declined by 37.17 million tonnes due to drought.

The State-wise details of the agricultural production in Kharif 2002 and estimates of damage, wherever available, presented below:

Table 5.8
Crop-wise Targets/Achievements

(Million tonnes)

Crop	*2000-01*		*2001-02*		*2002-03*	
	T	*A*	*T*	*A*	*T*	*A*
Rice	90.00	87.70	92.00	93.08	93.00	77.72
Wheat	74.00	69.68	78.00	71.81	78.00	68.69
Coarse cereals	33.00	31.08	33.00	33.94	33.00	25.10
Pulses	15.00	10.87	15.00	13.19	16.00	11.46
Total foodgrains	212.00	199.33	218.00	212.02	220.00	183.17
Nine oil seeds	28.00	18.44	28.00	20.73	27.00	15.44
Sugarcane	325.00	295.96	325.00	292.21	320.00	285.36
Cotton	14.50	9.52	14.50	11.69	15.00	8.94
Jute & Mesta	10.00	10.56	11.00	10.79	12.00	11.50

T – Target A – Achievement
Source: RBI, Reported on Currency and finance p. II-9.

Uttar Pradesh

All districts in Uttar Pradesh have been declared drought-hit. The crop-rich Western parts have been affected more badly. Poor irrigation infrastructure caused further damage to Kharif crop. The likely production estirhates for the State during kharif 2002 vis-a-vis targets are given on next page.

Haryana

On July 31, 2002 Haiyana was declared drought – hit with all the districts falling under deficient/scanty rain fall zone. Crops worst affected in this State include: Paddy, Bajra, Jawar, and Kharif pulses. It is reported that the Southern districts have been affected more than the Northern districts. Preliminary estimates worked out by the State Government indicate a total loss of around Rs. 665 crores, on account of

Table 5.9
Likely Production Estimated for Kharif 2002

	Area		*Production*	
	Target	*Achievement*	*Target*	*Achievement*
Rice	5860	3984	12900	6770
Coarse Cereals	2170	1427	3260	1250
Pulses	470	297	235	80
Total Food grains	8500	5708	16395	8100
Oil seeds	313	144	180	45
Sugarcane	-	-	-	-
Cotton*	20	9	6	3

Area in Thousand Hectares: Production in Thousand Tones
* Production in Thousand bales of 170.00 kg. each.

damage to crops and other additional expenditure, such as veterinary care, feed and fodder, etc. Around 9.6 lakh hectares area remained unsown during the kharif season due to drought.

Following relief measures have been undertaken by the State Government to provide immediate relief to farmers in the State.

- All the short-terms cooperative loans have been converted into medium-term loans.
- In case of farmers, whose kharif crop had suffered 50 per cent or more damage, as per special girdawari, the recovery of power bills for agricultural tube-wells for kharif shall be deferred by six months.
- It has been decided to remit the ablana for kharif 2002 immediately.
- An amount of Rs. 16 crore has been provided to Agriculture Department/Health Department and Animal Husbandry Department for providing

immediate drought relief services. Agriculture Department has been asked to provide the agricultural inputs, viz. seeds, pesticides, etc., to the farmers at 50 per cent price.

The likely production estimates for the State during kharif 2002 vis-a-vis targets as also the estimated damage are given below:

Table 5.10
Likely Production Estimated for Kharif 2002 and Estimated Damages

Crops	*Area*		*Production*		*Total Damage*
	Target	*Achievement*	*Target*	*Achievement*	*Estimated (in Rs. crore)*
Rice	950	745	2800	1720	90.00
Coarse cereals	790	414	910	120	97.06
Pulses	70	52	70	5	12.69
Total foodgrains	1810	1211	3780	1845	199.75
Oil seeds	6	3.7	5	2.3	0.92
Sugarcane	170	170	1000	-	-
Cotton*	610	550	1450	800	89.00

Area in Thousand Hectares Production in Thousand tonnes.
* *Production in Thousand bales of 170.00 kg. each.*

Punjab

In Punjab efforts have been made to save crops through canal or tube well irrigation's. This has an adverse effect on cost of cultivation. It is reported that most of the crop losses are confined to paddy and cotton. Preliminary estimates worked out by the State Government indicate the total loss of around Rs. 2,367 crore on account of damage to crops and other additional expenditure incurred on falling of water table levels and diesel. Around 1.7 lakh hectares of area remained unsown and 0.38 hectares under fodder got damaged during the kharif season due to drought. The following drought relief measures have been under taken by the State Government to provide immediate relief to the farmers in the State.

- Government has decided to provide relief of Rs. 700 per acre to the affected farmers as per special girdawari.
- All the deputy commissioners have been directed to ensure availability of drinking water in rural areas, wherever water level has gone down by installing hand-pumps immediately.
- Along with purchasing extra electricity, it was decided to divert power from urban/industrial sector to agriculture sector to meet the requirements of farmers.

Growth Rate of Agricultural Sector since 1950-51

Table 5.9 shows the overall growth rate in agriculture since 1950-51. It is clear from the table that during the first decade of planing, when the first and second five-year plans were implemented the annual rate of growth in agriculture was 3.3 per cent. But during the second and third decades of planning (1961-71 and 1971-81), the annual average of rate of growth in agriculture declined to 2.2 and 1.7 per cent respectively, mainly because of limited possibilities for area expansion. But, the conditions improved fourth decade of planning (1981-91). The sixth plan (1980-85) was grand success and the growth rate in agricultural production during this plan was 4.2 per cent and this contributed to a higher overall growth rate during the fourth decade. In the year 1991 the comply entered the new era of globalization and economic liberalization, which emphasis on commercialization of agricultural sector.

But, the growth rate achieved in the fifth decline (1992 to 2000) was not encouraging (only 2.2 per cent) This is mainly due to shift in the cropping pattern by the farmers towards short duration and high yield my varieties of

Table 5.11
Growth Rate of Agricultural Sector Since 1950-51

Period	*Compound Growth Rate (Percent)*
1951-61	3.3
1961-71	2.2
1971-81	1.7
1981-91	3.9
1951-91	2.8
1992-93 to 99-2000	2.2
2002-03 to 2001-02	3.2

Source: CMIE, Various issues and RBI Annual Report 2002-03 p. 268.

commercial crops. But in the year 2001-02-growth rate of agricultural sector has been marked the highest i.e. 5.7 per cent. This was mainly due to good monsoon, steadily increase in the use of agricultural inputs and implementation of special programmes like national water shed program for rain fed agriculture. But unfortunately the agricultural growth rate has been decline i.e. –3.2 per cent in the year 2002-03 due to drought.

As on August 2, 2002 as many as 50 of the 70 important reservoirs being monitored by the Central Water Commission had recorded storage of 30 per cent or less of the Full Reservoir Level (FRL)[5]. During July 2002, there was actually depletion in reservoir levels of as many as 24 of these reservoirs. There was actually depletion in reservoir levels of as many as 24 of these reservoirs. This painted a grim situation for the *Rabi* prospects. However, after the full monsoon season this grim situation has altered radically. Reservoir levels have looked up by the end of monsoon season. As explained in section 1.5 in some detail, the current level of 50 per cent of FRL for the 70 reservoirs is a radical improvement over the 18 per cent level that existed in end-July. The storage today is nearly 2.8 times the end-July level. However, this is only 2/3rd the average for the last 10 years period. The adverse impact on Rabi wheat could be minimized as reservoir levels in the three important wheat growing States, *viz,* U.P. Punjab and Haryana, which together account for nearly 3/4th of the country's wheat output, are good. There could be some adverse impact on wheat crop in Madhya Pradesh and Rajasthan on account of poor reservoir levels. Wheat production has ranged between 66-76 million tonnes over last year, averaging around 70 million tonnes. The target for 2002-03 been set at 78 million tonnes.

Kharif rice production for 2002-03 is estimated at 66.01 million tonnes as per the Quick Estimates. 16.8 per cent lower than last year. Important Kharif rice producing States like Uttar Pradesh, Andhra Pradesh, Punjab and Tamil Nadu had been hit by the dry spell in July. Rice production today is less dependent on Kharif than it was two decades back. Rabi/summer rice has become an important crop with states like Andhra Pradesh, West Bengal, Orissa, Tamil Nadu, Assam, Bihar, Kerala and Tripura contributing a sizeable production.

Coarse cereal production is less sensitive to surface and ground water availability than rice and wheat. Even otherwise, with rainfall deficiency and moisture stress in Maharashtra being low the prospects for jowar—the major coarse cereal crop can also be considered good, though it crop size production in neighbouring Karnataka could be affected at margin. Maharashtra accounted for 52 per cent and Karnataka for 21 per cent of All-India Jowar production in 2001-02. Official likely estimates for Kharif coarse cereals is place at 20.61 million tonnes, which is 23,8 per cent lower than last year and 5.2 million tonnes or 20.2 per cent less than the Kharif target for this year. However, the likelihood of Rabi coarse cereal production target of 7.16 million tonnes being attained this year is high. Under this scenario, total coarse cereal production would amountto 27.8 million tonnes, which will be nearly 7 million tonnes or about 20 per cent less than last year.

Madhya Pradesh is the most critical state for Rabi pulses as it accounts for 40-50 per cent of the total gram production in the country; gram being the major Rabi pulses crop. Uttar Pradesh, Rajasthan and Maharashtra are the other major gram producing states. The prospects for gram production can be considered reasonable with some possibility of average loss in Rajasthan and to a lesser Rajasthan and to a lesser degree in Madhya Pradesh as these crops are primarily grown by small and marginal farmers, whose income levels would have suffered as a result of Kharif crop failure and their existing entitlements would not be enough to support another sowing operation. Official estimates for the Kharif pulses place the crop at 4.23 million tonnes. Gram, crop can be expected to be higher than last year, but less than target, say at around 5.75 million tonnes. Total Rabi pulses output to be around 9 million tonnes; i.e., 1 million tonnes less than the target. Total pulses production in 2002-03 would then turn out to be around 13.2 million tonnes, compared with 13.5 million tonnes last year.

Regarding non-foodgrains, the likely estimates of the Government of India for Kharif 2002 at 117.2 lakh tonnes show a 11.2 per cent decline in the production of Kharif oil seeds, with groundnut, castor seed, seasmum and nigerseed registering large declines over last year's production levels. The decline in soyabean's production has been marginal, while sunflower production is higher by 6.2 per cent than last year. What news does this contain for our projections/ In absence of information, we had not done any detailed projections for non-foodgrains crop at the time of the August Report. This had become inevitable as we did not even have any area sown reports at that point of time. However, we now have considerable details on area shown for major oil seeds and Ministry's assessment for Kharif oil seed production. As such, it is possible to arrive at tentative estimates for oil seeds production for the full year. In 2001-02, Kharif oil seeds were placed at 132 lakh tonnes and Rabi oil seeds at 75.3 lakh tonnes, so that 63.7 per cent of the total production of nine oil seeds at 207.3 lakh tonnes for the full year came from Kharif crop and the balance 36.3 per cent came from Rabi crop. It may be mentioned that oil seeds production has exceeded 200 lakh tonnes mark for nine of the last 10 years, with a record production of 247.5 lakh tonnes in 1998-99. Rabi oilseeds had peaked at 99.8 million tonnes in 1996-97, but has since averaged 77.0 million tonnes in last five years (1997-98 to 2001-02). The likelihood of this average being bettered in 2002-03 is high on account of expected diversion in area to Rabi oilseeds crop in the aftermath of drought and the fact that, rapeseed and mustard crop can register some growth over the levels of 46.5 lakh tonnes in 2001-02 and 42.1 lakh tonnes in 2000-01 against an average of 61.6 lakh tonnes during preceding five years, viz., 1995-96 to 1999-2000. If rapeseed/mustard crop indeed turns out to be good, probably in the vicinity of the long-run average of about 60 lakh tonnes; Rabi groundnut turns out to be around 18 lakh tonnes in line with the average since 1990-91 and marginally higher than more recent averages; and other Rabi oilseeds also record small improvements over last year's level, total output of nine oilseeds could been the vicinity of 208 million tonnes, wiping out the Kharif losses completely so that oilseeds production for 2002-03 can be expected to be around last year's level. However, some downside risks in this projection remains and if rapeseed/mustard crop for instance of the type seen in last two years, total oilseeds production could drop by about 15 lakh tonnes over the preceding years total output of 207 lakh tonnes, i.e. by 7.2 per cent over last year's level.

Among other non-oilseeds crops cotton has been badly affected by drought. According to likely estimates of the Government of India, cotton production is slated decline by 25.38 per cent in 2002-03 to 88.7 lakh bales, from 116.9 lakh bales, due to acute drought conditions in Andhra Pradesh, Gujarat, Punjab and Haryana—the four states that together account for more than half of the total cotton crop in the country. Maharashtra, the major cotton producing state that account for about a fifth of total crop, however, received better rains. Jute and mesta output is expected to decline in 2002-03 by 7.9 per cent to 99.4 lakh bales from the previous year's level of 107.9 lakh bales. Sugarcane production, which generally is done under irrigated conditions, is also expected to fall for the second consecutive year to 282.58 million tonnes or 3.3 per cent from the previous year's level of 292.21 million tonnes.

It is clear from the above that drought has badly impacted agricultural, prospects for 2002-03. The loss of area coverage under different Kharif crops corresponding to the same period last year has been large. As on September 2, 2002 when sowing operations can be reasonably considered to have been over, the loss in area was as follows; rice about 7 million hectares (mha), coarse cereals 5.4 mha, oilseeds 1.3 mha, pulses 1.9 mha and cotton about 1.23 mha. The area coverage under rice in irrigated north-west states like Punjab was, however, almost achieved. However, the transplanted paddy crop has suffered due to delayed, erratic and subdued monsoon. There has been some withering of transplanted paddy in these states.

Following table 5.12 shows the Kharif foodgrain production in drought years.

Table 5.12
Kharif Foodgrain Production in Drought Years

Deficient Rainfall Years	*Monsoon Rainfall % departure from Normal*	*Rainfall in July (% drag)*	*Kharif food grain production (% decline)*
1972-73	-24	-31	-6.9
1974-75	-12	-4	-12.9
1979-80	-19	-16	-19
1982-83	-14	-23	-11.9
1986-87	-13	-14	-5.9
1987-88	-19	-29	-7.0
2002-03	-19	-49	-19.1

Source: DAC, Ministry of Agriculture.

Table 5.13

Index Number of Area, Production and Yield of Foodgrains, Non Food-grains and all Crops in India—growth rates in Drought Years

Year	*Foodgrains*			*Non-foodgrains*			*All crops*		
	A	*Pr*	*Y*	*A*	*Pr*	*Y*	*A*	*Pr*	*Y*
1	*2*	*3*	*4*	*5*	*6*	*7*	*8*	*9*	*10*
1972-73	-2.8	-8.1	-6.0	-5.8	-8.5	-6.4	-3.4	-8.2	-6.1
1974-75	-4.3	-5.0	-2.2	-0.5	-0.7	-0.7	-3.5	-2.9	-1.6
1979-80	-3.0	-18.2	-16.1	-5.3	-10.6	-7.8	-3.4	-15.5	-13.1
1982-83	-3.0	-3.6	-0.9	-2.2	-4.7	-4.8	-2.8	-4.0	-2.4
1986-87	-0.7	-5.3	-4.7	-3.4	-0.4	-0.0	-1.3	-3.6	-3.1
1987-88	-5.9	-2.9	2.2	1.0	5.2	1.8	-4.3	0.1	1.8
2001-02	-1.7	-5.1	-2.8	-2.4	-5.9	-3.1	1.9	-5.4	-2.9

A = Area *Pr = Production* *Y = yield*

Source: RBI, Handbook of statistics on the India on Economy 2002-03, p. 445

It is clear from the above table that, the decline in the rate of growth of foodgrain production and non-foodgrain production in all drought years.

In brief, number other drought in the post led to such a drop in food production as the 2002 drought, foodgrain production dipped by 29 million tonnes to 183 million tonnes i.e. 212 million tonnes in 2001. Over 18 million hectares of cropped area were left unsown during the kharif season. The per centage fall of kharif crop acreage, as compared to the normal, was the highest in Kerala (-59.3%) and Uttar Pradesh (-19.4%)

During the Rabi season, Rajasthan led the pack (-52.1%), as only 31.95 lakh hectares were sown against the normal of 66.69 lakh hectares. The other two states affect during the rabi season were Gujarat (-27.9%) and Tamil Nadu (-24.6%).

Drought of 2002 has been widespread and though largely limited to monsoon failure in July has badly affected agriculture prospects. Based on the initial reports for Kharif crop value added in agriculture could decline by 3.5 per cent, placing a drag on the overall growth rate that the economy can register in the current year. However, there are clear indications that industrial revival has been sustained and the sector could record an overall growth of 5.0-5.5 per cent in 2002-03. Services sector also remains buoyant with sustained acceleration in trade, hotels, transport and communications and financing, insurance, real estate and business services. Activity in other services also continue to be encouraging raising hopes that services sector as a whole could grow by nearly 8.0 per cent. As such, in spite 'of the severe droughts, Indian economy could still expand by 4.5 per cent or more, demonstrating that the medium-term growth prospects for the economy remain encouraging.

Impact on Industrial Sector

The slow growth of agricultural income impacts on consumer demand, which in turn affects the industrial output. Impact on private consumption is more pronounced in the case of consumer goods, which carry a weight of about 29 per cent in the index of industrial production. The supply side effect of the drought on the industrial output is transmitted through reduced supply of raw materials used in the Industries. The impact is two folds: one, the supply is reduced; second, the prices of such inputs for the industry tend to rise, thus raising the overall cost of production. The agro-based industries which constitute 20 per cent of valued added of the industrial sector, could be more severely affected by an adverse shock on agriculture since such industries entirely depend on the latter for requirements of their raw material. The impact can be broadly assessed from the fact that the raw materials in the agro-based industries constitute about 80 per cent of their total inputs.

Since the 1970s, Indian economy has witnessed a drop in agricultural production (negative growth) on as many as 13 occasions. Alongside industrial growth has had a corresponding decline on five occasions and responded with a lag of one year on five occasions. Only on three occasions, there had not been any coincident or lagged effect on the industrial growth. At the desegregated level, manufacturing growth has registered simultaneous decline on six occasions of drought and with a lag of one year for three drought years. It does not seem to have been affected in four drought years. Electricity growth too has recorded simultaneous decline for three drought years, decline with a lag of one year for eight drought years and remained unaffected in two drought years (Table 5.12).

Table 5.14

Impact of Drought on Industrial Growth (Since 1970s)

(No. of Years)

Growth	*Coincident Negative impact*	*Negative impact with 1 year lag*	*Total Negative Impact*	*No. Impact*
1	2	3	4(2+3)	5
1. Industry	5	5	10	3
(a) Manufacturing	6	3	9	4
(b) Electricity	3	8	11	2

The relation between industry and agriculture is possibly weak as it is an outcome that subsumes the impact on diverse constituents of industry such as mining and quarrying, electricity, gas and water supply, or construction. Significantly, manufacturing seemed to have been influenced more by the immediately preceding agricultural performance than by the current agricultural performance. At the desegregated level, the response of consumer goods to GVA (Gross Value Added) from agriculture was found to be relatively strong.

Drought impacts the performance of industry by influencing its supply channel. Such impact is predictably felt more in the case manufacturing and agro based industries in particular. For manufacturing and agro based industries, the relation between industrial raw materials and the GVA from agriculture was found to be statistically significant. Drought is, therefore, expected to affect the performance of both manufacturing and agro-based industries. Furthermore, in most of the drought years, there has been a negative growth in hydel power generation.

Table 5.15
Impact of Drought on Industrial Growth in the Industrially Advanced States (Since 1980-81 to 1997-98)

(No. of Years)

State	*Coincident Negative Impact*	*Negative impact with 1 year lag*	*Total Negative Impact*	*No. Impact*
1	2	3	4(2+3)	5
1. Andhra Pradesh	4	1	5	2
2. Gujarat	5	2	7	1
3. Maharashtra	3	2	5	2
4. Tamil Nadu	1	5	6	0

Table 5.16
Growth Rate of Major Industries During Past Droughts

Industry	*81-82*	*82-83*	*83-84*	*84-85*	*85-86*	*86-87*	*87-88*
Food Products	13.5	14.1	-6.5	-0.9	4.7	6.1	4.4
Beverages, tobacco and tobacco products	4.3	3.4	-3.1	6.9	0.4	-12.1	-13.8
Cotton textiles	-0.3	-10.3	12.1	2.0	8.0	1.9	-1.1
Jute, hemp and mesta textile	-4.3	-2.9	-15.8	27.1	-2.2	4.0	-9.9
Textile products (including weaning apparel)	-3.3	-0.4	-4.4	3.8	18.0	-22.8	5.3
Wood and wood products, furniture and fixtures	53.2	-0.1	9.5	29.3	3.1	10.3	-34.3
Paper and Paper products and printing industry	8.2	-2.5	3.6	20.7	12.5	10.0	1.9
Lether and fur products (except repairs)	28.1	-21.9	16.2	20.1	21.1	5.6	3.8
Rubber, Plastic, petroleum and coat products	19.1	-0.1	14.4	8.2	3.9	-2.2	3.9
Chemical and chemical products	16.9	3.7	8.1	9.0	8.1	13.7	14.5
Mon-metallic mineral products	6.7	-2.8	18.1	13.0	13.7	1.9	-1.4
Basic metal and alloy products	0.0	4.2	-8.7	12.8	9.0	8.4	6.9

Industry	*81-82*	*82-83*	*83-84*	*84-85*	*85-86*	*86-87*	*87-88*
Metal products & parts (except machinery & transport)	-5.4	-5.0	-2.0	19.2	9.2	8.5	4.2
Machinery and machine tools and parts	11.1	0.8	6.8	6.7	2.0	8.9	-1.8
Electrical machinery, apparatus and appliances	3.9	11.5	23.5	4.0	34.8	27.0	31.6
Transport equipment and parts	8.1	3.0	10.9	6.6	3.2	6.7	4.7
Other manufacturing industries	49.2	3.9	-32.5	17.4	24.3	54.2	15.6
MANUFACTURING (TOTAL)	7.9	1.4	5.7	8.0	9.7	9.3	7.9

Note- Shadings denote drought years.

As evident from the Table 5.13 manufacturing activities in these States (Andhra Pradesh, Gujarat, Maharashtra and Tamil Nadu) were adversely affected during the drought years. This could be attributed to the fact that the majority of the agro-based industries like food productsand beverages, cotton textiles, wool and silk textiles and textile products, etc are situated in these States, An analysis of industrial growth during past droughts show that agro-based industries like cotton, jute, hemp and mesta textile products, and those generating primary articles such as wood, paper etc., do gets adversely affected (Table 5.16)

Infrastructure Industries: A longer number of infrastructure industries achieved their production target set for 2002-03 than in 2001-02 than in 2001-02. Performance remained below targets in power, finished steel, crude petroleum, fertilizers and petroleum refinery products. The domestic civil aviation sector operated significantly below capacity. It is clear from the following Table 5.17.

Table 5.17
Targets and Achievements of Infrastructure Industries

Sector	*Unit*	*2002-03*			*2001-02*		
		T	*A*	*G(%)*	*T*	*A*	*G(%)*
1	*2*	*3*	*4*	*5*	*6*	*7*	*8*
Power		546	531	-2.6	540	515	-4.5
Coal	Million Tones	332	337	1.5	323	323	-0.1
Finished steel	Thousand T	36885	34377	-6.8	35079	31625	-9.8
Railway	M.T.	515	518	0.7	489	493	0.7
Shipping	M.T.	296	313	5.9	289	288	-0.5
Fertilizer	T.T.	16435	14440	-12.1	16589	14628	-11.8
Cement	M.T.	115	116	0.8	105	107	1.8
Crude Petroleum	M.T.	33.3	33.1	-0.6	33	32	-1.4

1	2	3	4	5	6	7	8
Petroleum & Petroleum products	M.T.	117	113	-3.7	114	107	-5.5
Civil aviation (Int.)	Lakh No.	112	123	9.9	112	117	4.5
Civil aviation (Domestic)	Lakh No.	215	170	-206	215	155	-27.9

Source: RBI, Annual Report 2002-03, p. 26.

Table 5.18
Index Number of Industrial Production Growth Rates

Year	*Quarrying*	*Manufacturing*	*Electricity*	*General*
1987-88	3.7	7.9	7.6	7.3
1999-2000	1.0	7.1	7.3	6.7
2000-01	2.8	5.3	4.0	4.9
2001-02	1.3	2.9	3.1	2.8

Source: Central statistical organization GOI.

It is clear from the above table that, general index number of industrial production is reducing.

Table 5.19
Index Number of Industrial Production Use Based Classification – Growth Rates

Year	*Basic goods*	*Capital goods*	*Intermediate Goods*	*Consumer Goods*	*Consumer durable*	*Consumer non-durable*
1	2	3	4	5	6	7
1977-88	5.5	15.9	2.7	9.8	7.6	10.3
1999-00	5.5	6.9	8.8	5.7	14.1	3.2
2000-01	3.6	1.8	4.7	8.0	14.5	5.8
2001-02	2.7	-3.4	1.6	6.0	11.5	4.1

Source: RBI, Handbook of statistics of the Indian Economy p. 447.

This table also showing the index number of industrial production in use based classification growth rates.

Annual Growth Rates of Major Sector Industries during past Droughts

The scenario as revealed by the growth of industrial production is one of a rising trend, with significant ups and downs over a number of years. But in the following table, growth rates are only major industries during past droughts.

Table 5.20
Annual Growth Rates of Major Industries during Past Droughts

Year	*Mining*	*Manufacturing*	*Electricity*	*General*
Weight	10.47 (11.46)	79.36 (72.11)	10.17 (11.43)	(100.00)
1981-82	17.7	7.9	10.2	9.3
1982-83	11.7	5.7	7.6	6.3
1985-86	6.2	9.4	10.3	9.1
1987-88	3.8	7.9	7.7	7.3
2000-01	2.8	5.3	4.0	5.0
2001-02	1.2	2.9	3.1	2.7
2002-03	5.8	6.0	3.2	5.7

Source: GOI, Economic Survey 1996-97, 1979-98, 1998-99, 1999-2000, 2003-04.

IMPACT ON SERVICES SECTOR

Services Sector Activity and its Linkages

The impact of drought on real activity in services sector is difficult to make in view of the weaker, not so quantifiable nature of the impact and the lack of information on actual activity and its lead indicators for this sector. It is, nevertheless, necessary to make some informed judgement about the real activity in services sector in spite of a relatively poorer understanding of the linkages of this sector with the rest of the sectors, as also the statistical problems associated with compilation of valued added in this sector.

It may be mentioned that the share of services sector in aggregate output was less than one-third in 1950-51. It increased over the years, consistent with the structural transformation of the economy. Importantly, the share has exceeded 50 per cent in each of the year since 1997-98, Sectoral shares are given in Table 5.21.

Table 5.21
Sector-wise Shares in Real GDP at Factor Cost

	Agriculture	*Industry*	*Services*	*GDP*
1950-51	57.4	10.6	31.9	100.00
1960-61	52.8	13.2	34.0	100.00
1970-71	46.2	15.5	38.2	100.00
1980-81	39.7	17.6	42.7	100.00
1990-91	32.2	23.7	46.1	100.00
1995-96	28.0	23.0	49.0	100.00
2000-01	24.2	22.1	53.7	100.00
2001-02	24.3	21.5	54.1	100.00

Note: Agriculture includes allied activities, viz. forestry and logging and fishing, Industry includes mining and quarrying, manufacturing and electricity, gas and water supply Services includes construction,' trade, hotels, transport and communication', financing Insurance, real estate and business services', and 'community, social and personal services'.

Source: CSO (2002).

Sub-Sectoral Performance in the Services Sector

Services regained their role as the principal driver of the Indian economy accounting for 89.4 per cent of the real GDP growth in 2002-03[6]. The resumption of near to end growth in services was enable by a pick up in activity in construction, financing, insurance, real estate and business services. This out weighted the deceleration in the growth of trade, hotels, transport and communication services, which constitute a little less than half of the services sector. It is showing in the following Table 5.22.

Table 5.22
Sub-Sectoral Performance in the Services Sector

Sub-Sector	*Share in Services Sector*		*Growth Rates*		*Relative contribution to growth of sendees GDP*	
	2002-03	2001-02	2002-03	2001-02	2002-03	2001-02
1	2	3	4	5	6	7
Trade Hotels TCC	43.3	43.0	7.8	8.7	47.3	56.7
Financing, Insurance real estate & business	22.6	22.8	6.1	4.5	19.5	16.2
Community, social & P.S. construction	24.7	24.7	6.8	5.6	23.7	21. 5
Construction	9.4	9.4	7.2	3.7	9.6	5.5
Services	100	100	7.1	6.5	100	100

Source: RBI, Annual Report, 2002-03, p. 30.

Table 5.23
Sector-wise Growth Rates and Droughts

	Agriculture	*Industry*	*Service*	*GDP*
1970-71	7.1	1.5	4.1	5.0
1971-72	-1.9	3.6	3.1	0.9
1972-73	-5.0	4.2	2.9	-0.5
1973-74	7.2	3.9	1.8	4.5
1974-75	-1.5	3.3	3.4	1.2

	Agriculture	*Industry*	*Service*	*GDP*
1975-76	12.9	4.2	7.8	9.4
1976-77	-5.8	8.4	5.3	0.9
1977-78	10.0	5.7	5.7	7.5
1978-79	2.3	11.2	5.4	5.1
1979-80	-12.8	-2.4	1.1	-5.2
1980-81	12.9	2.0	5.7	7.7
1981-82	5.3	8.8	5.4	6.0
1982-83	-0.7	7.3	4.5	2.9
1983-84	9.6	8.8	5.5	7.7
1984-85	1.5	6.4	5.2	4.0
1985-86	0.7	4.6	8.7	4.9
1986-87	-0.6	8.1	6.8	4.3
1987-88	-1.3	6.9	6.4	3.8
1988-89	15.5	9.8	7.3	6.5
1989-90	1.5	11.2	8.6	6.7
1990-91	4.1	6.8	6.1	5.6
1991-92	-1.5	-1.2	4.3	1.3
1992-93	5.8	4.1	5.1	5.1
1993-94	4.1	6.4	6.8	5.9
1994-95	5.0	11.3	6.9	7.3
1995-96	-0.9	12.9	10.0	7.3
1996-97	9.6	8.2	6.7	7.8
1997-98	-2.4	3.0	9.8	4.8
1998-99	7.1	2.8	8.0	6.6
1999-00	0.5	4.6	9.5	6.0
2000-01	-0.2	6.2	5.0	4.0
2001-02	5.7	2.9	6.2	5.4

Past drought years since 1971 have been shaded. In all these drought years, with the exception of 1985-86. All-India rainfall was less than 10 per cent of the Long-Term Average (LTA) or the per centage area affected by drought was 20 per cent or more.

Table 4.3 also shows that the Real GDP growth was negative in two of drought cases (1972-73 and 1979-80), while it was below 3 per cent in two other (1974-75 and 1982-83). So only in 4 out of 7 drought cases has real GDP growth fallen below 3 per cent.

Real activity that staged a smart recovery in 2001-02 after having slowed down in last two quarters of preceding year could be seriously affected by drought, 2002. Quarterly GDP data released by CSO on June 28, 2002 show that year-on-year real GDP growth that had dipped to 1.5 per cent in last quarter (Q-IV) of 2000-01 increased incessantly to 3.4 per cent (Q-I), 5.3 per cent (Q-II), 6.2 per cent (Q-III) and 6.4 per cent (Q-IV) in the four quarters of 2001-02 (Table 4.4). Quarterly GDP growth for Q-I 2002-03 is placed at 6.0 per cent, which though lower than the growth in preceding quarter is much above the 3.5 per cent growth of corresponding quarter last year. Rebound in activity in 2001-02 was largely supported by buoyancy in agriculture and allied activities, construction, 'trade, hotels, transport and communications', financing, insurance, real estate and business services'. Q-I 2002-03 results show that the acceleration in these sectors has been by and large sustained, though in case of 'agriculture and allied activities', as also construction, first quarter growth this year is lower than that of last quarter growth in preceding year. That the economy was at least clearly coming out of recession on the eve of monsoon failure. It obvious in that 'mining and quarrying' and manufacturing sectors recording acceleration for the third successive quarter; while 'trade, hotels, transport and communication' and 'financing, insurance, real estate and business services' recording an acceleration for the fifth successive quarter.

Some reassessment of the above can be made in light of the Q-I quick estimates of the CSO being available now. First, the construction sector recorded a growth rate of 6.3 per cent in 2002-03, which is over –0.2 per cent in the Q-I of 2001-02. As such, high Q-I growth in construction this year may not indicate much for the remaining three-quarters. This is important because construction growth in next three

Table 5.24
Quarterly GDP Growth Rates for 2000-01 and 2001-02

Industry Groups	*PERCENTAGE CHANGE OVER PREVIOUS YEAR*								
	2000-01				*2001-02*				*2002-03*
	Q-I	*Q-II*	*Q-III*	*Q-IV*	*Q-I*	*Q-II*	*Q-III*	*Q-IV*	*Q-I*
1. Agriculture, forestry & fishing	1.8	3.9	-0.8	-4.0	1.1	6.3	7.6	7.6	4.4
2. Mining & quarrying	4.8	3.6	4.3	0.9	-0.3	0.7	3.1	3.5	5.3
3. Manufacturing	8.1	7.1	7.1	4.6	2.7	2.6	2.9	3.1	3.8
4. Electricity, gas & water supply	7.1	4.5	9.3	4.0	3.9	5.4	3.8	5.4	5.3
5. Construction	12.4	10.0	7.2	-1.2	-0.2	2.7	4.4	7.5	6.3
6. Trade, hotels, transport & communications	8.1	6.3	4.8	2.4	4.5	6.3	6.6	7.2	7.4
7. Financing, insurance, real estate & business services	3.7	3.9	2.1	2.0	7.0	7.6	8.1	8.3	9.7
8. Community, social & personal services	3.2	9.8	5.0	5.8	6.5	5.4	6.2	5.6	5.6
9. GDP at factor cost	5.4	6.2	3.4	1.5	3.5	5.3	6.2	6.4	6.0

Source - C.S.O.

quarters were successively higher last year (see Table 4.3 of the next chapter) and this higher base effect could keep construction growth rates lower. However, there are some indications. The construction sector activity may be on the rise, because Quick Estimates of the Index of HP and Use-based Index released by the CSO. It is show that, capital goods industry for the period April-August this year has shown a growth of 7.3 per cent as against a negative growth of (-) 6.6 per cent in the corresponding period last year. According to the performance of six infrastructure industries being monitored by the Ministry of Commerce and Industry, infrastructure industries have shown marked acceleration this year with an overall growth of 6.7 per cent in April-August, 2002 compared to only 1.0 per cent in April-August 2001. These six industries have a high weight of 26.68 in the HP. Within the infrastructure, cement industry has recorded an impressive rise of 11.65 per cent in April-August 2002-03 compared with only 1.2 per cent in April-August 2001-02. Finished steel industry has also recorded significant acceleration with a growth of 8.9 per cent in April-August 2002 compared with a negative growth of (-) 0.9 per cent in April-August 2001. of 6.1 per cent that would be in largely line with the steady growth of similar order observed for sometime now.

However, the Q-I GDP estimates of CSO show exceptional growth of 9.7 per cent in financing, insurance and business services, up from 7.0 per cent last year. While it may not be possible to sustain these high growth rates for next quarter, with some drought affect coming in this period, the prospects for this sub-sector remain quite good in view of the industrial recovery that has been pronounced in the Q-III of 2002-03. The growth rate that may be recorded in next three-quarters of 2002-03 may be somewhat lower than that in Q-I. As the higher base effect may come into play with financing, insurance, real estate and business services having already recorded a pronounced acceleration last year. The growth rates in each of the four quarters of 2001-02 being markedly higher than any quarter of 2000-01. Also, growth in each quarter of 2001-02 was successively higher than growth rate of the preceding quarter of the same year. This sort of acceleration would put some drag on the growth of this segment in Q-II to Q-IV of 2002-03. It may, nevertheless, still be possible for this sector to record a growth of 8.0-9.3 per cent in full 2002-03, with a central projection of 8.5 per cent. The upward revision in our central projection is on account of higher than expected Q-I growth and the renewed activity in industrial sector.

GDP Growth in Past Droughts

An analysis of real GDP growth since 1970-71 reveals that in case of 6 out of 7 drought years in the 1970s and 1980s valued added in agriculture and allied activities, declined, with 1985-86 being the only exception (Table 5.25). Deceleration in valued added in industry in the year of drought or its succeeding year was generally witnessed. The deceleration was contemporaneous in 5 cases though the causation is not very clear. For instance, the negative growth in 1979-80 could have been related to the OPEC induced second oil price shock. The impact on service sector is also not easily decipherable. However, real GDP growth fell in each of the drought years, excepting 1985-86 where rainfall was only 7.1 per cent less than LTA though a large area 32.2 per cent of the total got drought affected. In that sense 1985-86 does not strictly qualify as meteorological drought year, though it has been identified as one by the Government of India Report of 1989 due mainly to adverse temporal and spatial factors of rainfall in that year. Furthermore, since 1990-91, agricultural growth has been negative in three years, implying that the probability of negative growth in agriculture is still a possibility in spite of technology advances. So for all practical purposes, a negative growth in valued added in agriculture and allied activities for 2002-03 can be anticipated as a stylized outcome on the basis of past data. The question is what would be the size of the fall.

Impact on Hydroelectric Power

The impact of the drought 2002 on hydroelectric power generation led to a decline of 13.9 per cent. The per centage reduction in power generation is the maximum when compared with the drought years of the recent pas. These are in the following table.

Table 5.25
Impact of Drought on Hydroelectric Power Generation

Droughty Year	*Expected level*	*Actual*	*Loss*	*% Loss*
1979-80	47.1	45.5	1.6	3.4
1982-83	49.6	48.4	1.2	2.4
1987-88	53.9	47.5	6.4	11.9
2002-03	73.9	63.7	10.3	13.9

Source: Ministry of Power/Ministry of Agriculture.

Impact on Capital Market

The impact of Drough on stock markets was already substantially discounted, though uncertainties exit over the feature course of the stock prices. The stock prices slump during July 9, August 8, 2002, substantially on account drought with SENSEX as well as Nifty dropping by about 12 per cent. Stock prices, however, rebounded subsequently with nearly 8 per cent gain by early September over the laws of August 8, 2002. These gain however, have dissipated with the SENSEX and Nifty hovering at around the August 8 low levels. The weakness over the last month or so was caused by factors other than drought in which political events, including uncertainty on disinvestments policy, and sovereign credit rating down grade had a role to play.

The FMCG sector, which is partly dependent on rural demand, was one of the worst hit, with the FMCG Index declining by 14.0 per cent during 2002-03. Among the Sensex script, FMCG corporate like HLL were adversely affected, declining by 21.7 per cent during the period end-March to

September 26, 2002. Certain other sectors, such as fertilizers, which are equally dependent on agricultural demand, however, did better, as sectoral valuations were boosted by hopes of Government disinvestment in the sector. The SandP CNX Food and Food Processing Index registered a sharp decline of 3.2 per cent during the period end-March to end-July 2002 but recovered thereafter.

The equity markets continue to be weak, with the BSE Sensex declining to an average of 3123.4 during the second quarter of 2002-03 from 3334.0 during the first quarter of the year. While the drought continues to weigh on sentiment at the margin, it does appear that the markets have already factored in the resultant adverse impact on rural demand, Besides, the news of industrial recovery has, to some extent, counter balanced the negative impact of the expected fall in agricultural output. By the latest indications, it is likely that the markets would remain range-bound in the near term, especially in view of the downside risks of uncertainties regarding PSU disinvestment and the continuing weakening of international markets.

Impact on India's Foreign Trade

In order to comprehend the total picture of trade, as also it's performance, we take up first the significant trends in the value or trade as a whole, as also in it's two important components, namely, imports and exports. It is clear form the following two Tables 5.26 and 5.27.

Table 5.26
Impact on Foreign Trade

Year	*Imports*	*Exports*	*Total value of trade*	*Balance of trade*
1999-00	2153226	159561	374887	-55765
2000-01	230873	203571	434444	-27302
2001-02	245200	-209018	454218	-36182
2002-03	297206	255137	552343	-42069

Source: GOI, Economic survey 2004-05.

Table 5.27
Exports in Major Industry Group

Major Industry group	*No. of reserved items as on May 2003*	*Export growth (percentage)*	
		1991-95	*1995-2002*
Mechanical Engineering	183	12.2	11.1
Rubber, glass, paints, elements products	48	19.3	7.5
Plastic products	13	43.1	12.5
Chemicals, organic chemicals, drugs & plasma	113	11.7	11.4
Sports goods	7	10	0.8
Food & Allied Products	12	18.7	4.5

Source: RBI, Report on Currency and Finance 2002-03, p. 93.

Fiscal and Monetary Situation

The failure of monsoon during the current year has brought a new uncertainly to economic outlook, which so far had seemed to indicate a recovery. In order to gate some idea about the extent of possible damages in a bad agri-growth year, we looked at the historical data for past sixteen years. The data suggest that poor agri-growth years have typically had a larger impact on the growth rates than on inflation. During 1986-87 to 2001-02, there was a drop in agriculture GDP for six years.

However, the impact was more on the foodgrains outputs in these years. Particularly on Kharif crop. During these six years, the Kharif foodgrains output dropped on an average by 5.2 per cent. The average drop in foodgrains output at 4.5 per cent was some what moderated by the rabi foodgrains, whose output dropped by 3.3 per cent on an average. But the average drop in agricultural GDP in these years at 1.2 per cent was much lesser than in foodgrains, reflecting the diversification into how grain crops. 1986-87 was the worst year for the sector, when agricultural GDP had declined by 2.4 per cent.

Going by the past experience, it are expects a drop of 1.2 per cent in agri-GDP in 2002-03, then it would lead to a direct drop of 0.3 to 0.5 per cent in overall GDP. In the lost fiscal agricultural had contributed 1.4 per cent to GDP growth. A reversal from that position would mean, in an incremental sense, a drop of 1.5 to 2 per centage points in the economics growth rate.

A part from this direct impact, there could be some indirect impact on growth too, to the extent that the rest of the economy depends on rural demand. In the past, the performance of the non-agricultural GDP has not been distinctly different in the good and bad agricultural years.

Economic slowdown as a result of widespread drought conditions in the country is expected to put considerable stress on the finances of the Government. While Government would be required to undertake additional expenditure on account of extensive relief programmes, its revenue collection may fall short of the targeted level owing to the probable slowdown in both agricultural and industrial output. Assuming that no additional revenue mobilization measures would be taken, imminent draught is likely to widen the fiscal gap considerably. This section attempts an estimate of the likely rise in fiscal deficiton account of drought conditions as a result of fall in revenue and increased expenditure.

Impact on Expenditures and Relief Funds

It is the primary responsibility of the States to incur necessary expenditure on the immediate relief whenever a natural calamity occurs. The role of the Center is to provide supplementary assistance to the States, as it may not be possible for a State to immediately come forward with sufficient funds to meet natural calamities which occur suddenly and with intensity. As per the present provisions/ arrangements, for providing assistance to states, resources could be tapped through two special funds set up for undertaking natural calamity relief operations. These are: Calamity Relief Fund (CRF) and the National Calamity Contingency Fund (CRF).

Calamity Relief Fund (CRF)

First claim could be made on CRF, which is permanent in nature and resources are contributed to this fund as per the recommendations of the Finance Commission. As per the Eleventh Finance Commission, corpus of CRF for the year 2002-03 is around Rs. 2100 crore.

National Calamity Contingency Fund (NCCF)

Since 2000-01, NCCF scheme has been put in place to be activated in case of calamity of severe nature, where States are not in position to provide adequate relief from, the CRF and its own resources. As per norms, any assistance provided by the Center from NCCF should be financed by levy of a special surcharge on the Center taxes for a limited period.

Criterion for Determining Level of Assistance under CRF and NCCF

The Eleventh Finance Commission has indicated the following criteria for determining the assistance from the Center to the states available under CRF and NCCF.

For drought conditions, commission has recommended that small and marginal farmers may be provided assistance as below:

For crops which experience a loss of 50 per cent or more, should be provided assistance (as input subsidy) at the rate of Rs. 1,000 per hectare in case of rain fed areas and Rs. 6 per day per animal for small animals.

Animal husbandry assistance is also provided under other heads such as procurement and storage of fodder, movement of useful cattle to other areas, etc. Apart from these, relief could also be provided for items such as employment generation, supplementary nutrition, supply of drinking water. However, for most of these items there are no pre-determined norms and the requirement is assessed by National Center for Calamity Management (NCCM) team.

The extent of drought relief of assistance from Center to the states, as per above norms, could be estimated only for two items viz., crop loss and fodder for livestock as there are no predetermined norms for other items. As per above norms out estimates for additional expenditure or assistance under CRF/NCCF, works out to Rs. 2,700 crore for crop per day in case of live stock. These estimates pertain to 13 drought affected states viz., Andhar Pradesh, Chhattisgarh, Haryana, Himachal Pradesh, Karnataka, Kerala, Madhya Pradesh, Gujarat, Punjab, Rajasthan, Tamil Nadu and Uttar Pradesh. Table 5.28 indicate that extent of crop loss irrigated areas is less than 50 per cent. Since relief is available only in case of drop loss of 50 per cent or more, estimates for additional expenditure on account of crop loss has been worked out on the basis of rain fed areas only.

Table 5.28
Relief for Small and Marginal Farmers in the Drought Affected Areas

	Irrigated	*Rainfed/ Unirrigated*
Total Area under principal crops (in million hectares)	63	93
Of which land in 13 affected States (in million hectares)	53	82
Of which belonging to small and marginal farmers (in million hectares)@	22	27
Relief (Rs. Crore} $	0	2700

As indicated above, apart from relief on account of crop loss and affected livestock, States are also required to incur additional expenditure for employment generation program, supplementary nutrition, supply of drinking water, etc. Since there is no pre-fixed criterion for ascertaining the quantum of relief under these heads, it would not be possible to determine the overall quantum, of additional expenditure for drought relief, using the criterion laid down by the Finance Commission.

Animal Husbandry Assistance

Assistance under this item has been worked for 13 states supposed to the drought affected. Size of livestock for these states was worked out separately for the large and small animals by using the proportion of each state in All India milk and wool production, respectively. Further, the livestock eligible for drought relief has been estimated as below:

Table 5.29
Relief for Livestock in the Drought Affected Areas

	Large	*Small*
Total Livestock in 13 States (in 000)*	223931	136231
Of which belonging to small and marginal farmers(in 000) @	74644	45410
of which living in rainfed areas (in 000) #	44786	27246
Relief (Rs. Crore) per day	54	16

* Assumed to be 80 per cent of the total based on the share of these States in total milk and wool production.

@ Assumed to be one third based on the share of small and marginal holdings in the total land holdings on all India basis.

\# Assumed to be 60 per cent based on the ratio of rain fed areas in these States to the total Area.

According to official data, the Drought Management cell in department of agriculture and Co-operation is monitoring drought situation in Andhra Pradesh, Chattisgarh, Haryana, Himachal Pradesh, Karnataka, Madhya Pradesh; Maharashtra, Orissa, Punjab, Rajasthan, Tamil Nadu and Uttar Pradesh. Detailed information in respect of states having declared drought-hit available with drought Management Cell in Department of Agriculture and Co-operation is furnished in Table 5.30.

Table 5.30
Central Assistance Required

State	*Districts/Area declared drought affect (Nos)*	*Assistance requested from National Calamity Contingency Fund (In Rs. crore)*	*Foodgrains demanded (in lakh million tonnes)*
Andhra Pradesh	22 (836 Mandals)	1210.90	25.00
Chhattishgarh	16 (98 Tahsils)	2193,96	12.60
Gujarat	NR	NR	NR
Haryana*	19	1985.98	9.72
Himachal Pradesh	12	155.86	0.19
Jammu & Kashmir	NR	1623.98	NR
Jharkhand	22	1467.25	NR
Karnataka	24 (152 Talukas)	553.52	6.00
Kerala*	NR	NR	NR
Madhya Pradesh	31 (176Tehsils)	819.62	2.50
Maharashtra	33 (353 Talukas)	1730.61**	2.00
Orissa	30 (283 Blocks)	871.40	12.19
Punjab	17	3176.48	NR
Rajasthan	32	7519.76	56.00
Tamil Nadu	NR	720.00	2.00
Uttar Pradesh	70	7539.79	20.00
Uttaranchal	13	401.81	NR
West Bengal	3	NR	NR

NR: Not Reported; * Under examination

** Requirement of Maharashtra includes reimbursement of excess expenditure of Rs. 530.61 crore incurred during last two years.

Alternative Estimates for Additional Expenditure based on Past Experience

Since no precise estimates about the additional fund

requirement for the drought relief could be made from, the available information so far, an alternative approach could be to use the trends in expenditure during the period when similar drought conditions had occurred in the past. However, the information available in the Union Budget on transfers from Center to the states under loans and advances/ grants-in-aid', would only be useful in this regard as it is not feasible to segregate thee expenditure incurred by States for drought relief from other normal expenditure. In this context, Eleventh Finance Commission (EFC), has observed that: It is seen from the expenditure data of the States that it is very difficult to distinguish between the expenditure incurred for calamity relief and other normal expenditure under various heads of account.

Union budget for the year 1988-89 states that during the year 1987-88, an amount of Rs. 2,000 crore were transferred to states towards drought relief measures. This additional allocation Rs. 2,000 crore to the states work out to around 3 per cent of the total expenditure of Union Government in that year. Assuming that the drought conditions in the current year are similar to those in the year 1987-88, as per the initial estimates based on rain-fall data during the month of July 2002, the crop loss on all-India basis was expected to be much higher than during the year 1987-88. However, conditions seem to have improved somewhat with improved rainfall in subsequent months. Therefore, it could be presumed that over all crops loss during the current financial year would be of the same order as in 1987-88. Center may be required to earmark 3 per cent of the total expenditure in the current year. Accordingly, additional gross allocation of the order of Rs. 12,000 crore could be made by the center towards meeting the States requirements for drought relief measures.

Union Budget had already made a provision of about Rs. 1700 crore towards CRF and Rs. 1600 crore towards NCCF. Moreover, on account of different welfare schemes including the food for work programme, which are likely to be the part of drought relief measures, the size of buffer stock would get reduced. The reduced buffer stock would result in savings of around Rs. 2900 crore in carrying cost. Considering the provisions made for calamity funds and likely savings on account of buffer stocks operations, net additional expenditure Which need to be made by the Center towards drought relief would be. of the order of around Rs. 5,800 crore.

Additional Expenditure by State Governments

As per the norms, State Governments are required to contribute to drought relief from its own sources to the extent of 33 per cent of the assistance received from the Central Government. Therefore, as per the above estimates of additional expenditure by the Union Government of the order of Rs. 12,000 crore, State Governments would have to make contribution of Rs. 4,000 crore towards drought relief. However, net additional expenditure excluding the states share in CRF and NCCF (provision already made) amounting to Rs. 550 crore and Rs. 530 crore, would work out to Rs. 2920 crore.

The likely share of each State in the total additional expenditure (States own contribution) has been indicated in the Table 5.31 below. State-wise share has been arrived at on the basis of the relative size of each State's total expenditure budgeted for the year 2002-03.

Table 5.31
Increase in Expenditure – Drought Affected States

(Rs. Crore)

	Total Expenditure 2002-03 BE	*Expected Addition to (2)*
1	2	3
1. Andhra Pradesh	35420	360
2. Chhattiishgarh	7057	72
3. Haryana	11961	122
4. Himachal Pradesh	6014	61

1	2	3
5. Karnataka	25699	261
6. Madhya Pradesh	18140	184
7. Gujarat	29225	297
8. Orissa	14165	144
9. Punjab	19178	195
10. Rajasthan	22420	228
11. Tamil Nadu	30831	313
12. Uttar Pradesh	51700	526
13. Kerala	15360	156
Total for 13 States	287170	2920

Impact on Revenue

Generally, fall in agricultural output on account of drought conditions is likely to have two pronged impact on the economy, i.e., fall in overall output or real GDP and rise in price level. A priori, government tax revenues are expected to be positively related to each of these variables. While fall in output would raise the revenue collection.

The likely fall in revenue collection of each State has been indicated in the Table 5.32. State wise details have been worked out on the basis the relative size of each State's total own tax revenue budgeted for the year 2002-03.

Additional Deficit

On the basis of revenue shortfall and additional expenditure which would be required to be incurred on account of drought relief measures, the additional deficit for Center and States Governments have been estimated in above Table 5.31.

Table 5.32

Fall in Revenue Collection – Drought Affected States

(Rs. in crore)

	Own tax rev 2002-03 BE	*Expected short fall in (2)*
1	2	3
1. Andhra Pradesh	13460	591
2. Chhattishgarh	2130	94
3. Haryana	5549	244
4. Himachal Pradesh	888	39
5. Karnataka	11887	522
6. Madhya Pradesh	5661	249
7. Gujarat	14625	642
8. Orissa	2880	126
9. Punjab	6423	282
10. Rajasthan	7028	309
11. Tamil Nadu	14254	626
12. Uttar Pradesh	134484	592
13. Kerala	8709	382
Total for 13 States	106979	4696

Table 5.33

Calculations of Additional Deficit of the Central Government

Central Government

(Rs. Crore)

I	Fall in Net Tax Revenue	4710
II	Net additional expenditure (1-2-3)	5800
	1. Additional expenditure	12000
	2. Provisions in CRF / NCCF	3300
	3. *Savings due to Buffer Stock*	2900
III	Additional Deficit (I+II)	10500

State Governments

		(Rs. Crore)
I	Fall in Revenue (1+2)	6610
	1. *States' own Tax Revenue*	4780
	2. *Share in Central Taxes*	1830
II	Net additional expenditure (1-2)	2920
	1. *Additional expenditure*	4000
	2. Provisions in CRF/ NCCF	1080
III	Additional Deficit (I+II)	9530

Combating Drought with Prudent Fiscal Management

Central Government

The options available to the Central Government to meet the additional expenditure on drought relief are as follows:

1. If the extent of drought covering 13 States is considered to be grove enough to term it as a national calamity, the NCCF provision allows the Union Government to levy an additional surcharge. Given the additional requirement of Rs. 10,500 Crore, a surcharge of about 4.5 per cent on all central taxes would need to be imposed. Experience of earlier years also shows that a part of the additional expenditure on account of natural calamities was finance through levying of special tax surcharge e.g. tax surcharge in case of Gujarat Earthquake in 2000-01 and drought in 1987-88.

2. However, if no measure for additional revenue mobilization are taken, it is expected that fiscal deficit for the Central Government would increase by Rs. 10,500 crore over the budgeted level. Accordingly, the GFD of the Union Government as a ratio to GDP would rise to 5.7 per cent as against the budgeted level of 5.3 per cent.

3. Additional borrowings by the Central Government to finance the increased fiscal deficit would result in rise in debt- GDP ratio to 59.3 per cent as against the budgeted level of 59,2 per cent.

State Governments

Options available to the State Governments to meet the additional expenditure are relatively limited.

1. State Governments have limited avenues for additional resource mobilization, as all major taxes except sales tax, are collected by the Central Government.

2. Similarly, under the existing borrowing arrangements, the States cannot unilaterally take a decision to increase its market borrowings. The market borrowings programmed of State Governments are finalized by the Government of India in consultation with the Planning Commission. During the last few years, the States borrowing program has come under pressure. States have responded by resorting to non-transparent practices of raising funds outside the market-borrowing programmed, such as, debt securities issuance sponsored by State Governments, funds raised by backing of States Government guarantees, as also those raised through Special Purpose Vehicles (SPVs).

Additional expenditure (States' own contribution) towards drought relief by the states would probably require the corresponding increase in borrowing to meet this expenditure. This is because, all the States are already running deficits and prospects for generating additional revenue are rather bleak.

Thus, assuming that no additional revenue generation takes place, the drought related expenditure by States (States own contribution), aggregate gross fiscal deficit of the states as a ratio to GDP would rise from 3.8 per cent to 4.2. Correspondingly, the debt-GDP ratio of the states would increase to 26.9 per cent from the budgeted level of 26.5 per cent during year 2002-03.

In addition to the above, with a view to manage short-term liquidity mismatches, States can access Ways and Means Advances (WMA) from Reserve Bank. At present, the normal limits of WMA for each state has been fixed by the Reserve Bank on the basis of the average of revenue receipts and capital expenditure during the three years from 1999/99 to 2000/01. Special WMA limits are fixed on the basis of the

Table 5.34
WMA and Overdraft Position - Drought affected States

(Rs. Crore)

	Limits on WMA					*Additional*
	Normal	*Special*	*Total*	*Outstanding*	*Overdraft*	*Expenditure*
1	2	3	4	5	6	7
1 Andhra Pradesh	520	42	562	0	0	951
2 Chhattishgarh	100	363	463	0	0	165
3 Haryana	180	11	191	0	0	365
4 Himachal Pradesh	115	0	115	115	8	100
5 Karnataka	375	23	398	0	0	783
6 Madhya Pradesh	275	20	295	295	106	433
7 Gujarat	445	16	461	0	0	939
8 Orissa	185	18	203	203	18	270
9 Punjab	235	20	255	233	0	477
10 Rajasthan	310	156	466	440	0	536
11 Tamil Nadu	415	78	493	450	0	939
12 Uttar Pradesh	630	237	867	0	0	1118
13 Karala	225	12	237	237	145	539
Total for 13 States	4010	996	5006	1973	277	7616

State's outstanding- investments in Central Government securities. Over and above these (normal plus special), States are allowed to run an overdraft for 12 consecutive days.

Given the limited resource at the disposal of the State Governments and the need for drought relief measures, it is likely that the drought-affected states would be utilizing WMA/ overdraft facility more aggressively. Moreover, those states, which are facing relatively worse drought conditions or are having fragile fiscal situation, may demand temporary relaxation of WMA limits. The latest daily position of WMA and overdrafts as on October 3, 2002 reveals that out of the drought affected States, Himachal Pradesh, Madhya Pradesh, Orissa and Kerala have already exhausted their WMA limits and are into overdrafts. Table 5.34 below indicates the state-wise details of eligibility's under WMA/ over draft scheme, current level of utilization vis-a-vis additional expenditure requirements.

To conclude, the impending drought conditions in several parts of the country are expected to put further pressure on the already fragile finances of the Government. This equires the Government to bring in more discipline, economy and efficiency in the conduct of their expenditure, while mobilization of additional resources through better tax administration, if the negative impact on deficit and public debt is to be brought down. However, the alternative option of imposing surcharge to make up the increased fiscal gap could be avoided in the light of following facts about the current economic conditions:

- Economy is reeling under recession for the past two years and only recently in the current year, there are some signs of revival.
- Inflation is at a historic low level.
- Interest rates are lowest in the past 20 years.
- Country has the comfortable for exreserves position,
- Huge food buffer stock is available.

References

1. India Meteorological Department.
2. R.B.I. Annual Report 2001-02.
3. RBI, Report on currency and finance 2002-03 p. 10.
4. Source-RBI, Report on currency and finance 2001-02 p. 7.
5. RBI, Report on currency and Finance 2001-02 P. II. 7.
6. RBI, Annual Report 2002-03 p. 30.

6

Drought Relief Measures

The 2002 drought, however, saw a departure from the past in terms of setting up a response mechanism. It is being touted as a runaway success for its efficacy. As soon as the CWWG hit the trigger, on July 16, the Prime Minister set up a 'task-force on drought management' under the chairmanship of the deputy Prime Minister. Its members included ministers of agriculture, finance, food, rural development and the deputy chairman of the Planning Commission.

As the apex body, the task force not only took policy decisions but also reviewed the progress of relief operations. The NCMC merely functioned as a committee of secretaries to chew on issued referred to it by the task force. The apex body enforced a series of measures — from the release of funds to contingency crop plans, drinking water supply, provision of fodder and food for work programmes.

The process for releasing funds to the states, under the National Calamity Contingency Fund (NCCF) also become easy, as the constitution of the task force (except for ministers for food and rural development) was similar to that of the High Level Committee (HLC) that normally sanctions financial assistance. The task force therefore doubled as the HLC.

To do away with procedural delays, the task force was

empowered under Rule 12 of the Government of India (transaction of business) rules. Its decisions were thus deemed to be a decision of the cabinet.

There is no common or approved pattern for the states. The relief minister or the revenue minister heads the system and the coordination of the entire operation is usually in the hands of the states chief secretary. Many states have a relief commissioner heading the relief department.

As the district level, the collector is in charge of drought/disaster management and is assisted by officials from the revenue and other departments. In some states, officials in the relief department are posted to the district and work under the collector.

Under the present doctrine, the primary responsibility of extending financial support to the states to cope with disasters rests with the central government. Over 80% of the total money spent on relief by the State Governments come from the Center.

The process of sanctions from the Calamity Relief Fund (CRF and the NCCF is set in motion with a formal declaration of drought by the affected states. It is done on the basis of enabling legislation, codes or manuals and an assessment of damage to crops by undertaking crop-cutting experiments. The drought of 2002, caused by the failure of the monsoons in July, did not allow for such experiments, as sowing in many areas had not been carried out (110 lakh hectares across the states). Therefore, drought was declared on the basis of 'eye-estimation'.

Once a drought is officially declare, the states start submitting memoranda assistance from the NCCF. A central team/s (usually headed by an officiate of joint secretary) then fans out to the affected states to assess the situation ground and deliberate with the state government. An Inter-Ministerial department headed by the agriculture secretary scrutinizes the recommendations of central team.

The IMG applies the norms of expenditure laid down for the CRF and moves the scrutiny report to the final authority, the HLC. This time around had to be relaxed.

Usually, agriculture input subsidy is payable to small farmers when their crop is damaged to an extent of 50% or more due to this.

As large areas remained unsown, it was decided that the area normally under kharif cultivation by considered the benchmark. Farmers owning two hectares of land were also made eligible.

The subsidy rates were Rs. 1,0000, Rs. 4, 000 per hectare for irrigated and perennial crops respectively. The rationale for the input subsidies compensate farmers for losses incurred but to provide ex-gratia payment farmers and their families cope with the situation and help them organ crop. The compensation aspect is handled by crop insurance schemes. The flow from the Center to the states has been Rs. 12,950.95 crore for drought which included the value of foodgrain released (Rs. 8736.00 crore).

Total Assistance from the CRF (Central Share) and NCCF Released to the States.

(value of foodgrain also included (ruppes in crore; foodgrain in MT)

State	*Central share of CRF*	*Assistance from* NCCF
Andhra Pradesh	249.75	123.51
Chhattishgarh	34.64	127.51
Gujarat	133.46	–
Haryana	102.52	–
Himachal Pradesh	35.96	14.35
Jammu & Kashmir	28.86	–
Jharkhand	71.49	–
Karnataka	94.03	207.65
Kerala	84.79	–
Madhya Pradesh	78.97	171.66
Maharashtra	198.23	20.00

State	*Central share of CRF*	*Assistance from* NCCF
Orissa	138.04	5.29
Punjab	101.47	–
Rajasthan	350.88	889.61
Tamil Nadu	84.87	332.09
Uttarakhand	40.81	–
Uttar Pradesh	184.45	310.06
Total	2013.22	2201.73

Source: Ministry of Agriculture.

Almost half the allocations made by the HLC under the NCCF went as input subsidies (49.92%) to farmers, with cattle care (19.30%) and employment generation (10.78% taking major shares. The break-up for drought 2002-03 is in the following:

Item	*Percentage of total amount sanctioned*
Input subsidy	49.92
Cattle care	19.30
Drinking water supply—rural	9.75
Drinking water supply – urban	4.80
Employment generation	10.78
Gratuitous relief	1.41
Supplementary nutrition	3.11
Health	0.39
Others	0.54
Total	100.00

- The per centage of relief under employment seems small, as generation of employment was undertaken largely under the food-for-work regime.
- The relatively small expenditure under gratuitous relief, as compared to previous droughts, is due to massive relief employment programmes undertaken during drought 2002-03.
- The allocation under water supply excludes expenditure under the 'Emergency' component of the Accelerated Rural Water Supply Programme.
- Expenditure by the railways (Rs. 44.64 crore) for free transportation of water and fodder to Gujarat, Rajasthan and Andhra Pradesh has not been included. The sanction for cattle care does not include an amount of Rs. 4.69 crore released by the Prime Minister's Relief Fund for goshalas in Rajasthan.

Crop Insurance

The NCF and NCCF funds are largely provided that to rural populations to tide over the pauperizing conditions created by drought and to equip them with the wherewithal to prepare for the ensuing sowing season.

Compensation for loss of crops during a drought season is a function of the National Agricultural Insurance Scheme. The newly formed Agricultural Insurance Company of India (AIC) is experimenting with a number of innovative schemes that factor in rainfall deficiency. Varsha Bima, for instance, kicks in when rainfall is 20% below normal. The claim pay out to farmers is on a graded slab (rupees per mm of rainfall). The scheme has been tried out in 20 rain gauge areas across the nation.

The NAIS settled claims of over Rs. 87,850 lakh for kharif 2002, in the states shown in the table. Total claims paid out for the rabi season were over Rs. 9,132 lakh. However, the NAIS scheme is far from perfect as claims are settled only on the basis sought to lessen the burden of farmers in distress by taking a number of measures including:

- Refraining from the recovery of principal or interest on kharif loans during 2002-03.
- Conversion of kharif crop loans to medium-term loans payable over five years.
- Waivers on interest on pending crop loans during 2002-03; no penal interest on individual agricultural loans up to Rs. 25,000 borrowed during calamities.
- Release of fresh loans to farmers.
- NABARD issued guidelines to cooperative banks for investment credit installation of tube wells, purchase of fodder.

The drought of 2002-03 consumed over Rs. 6,000 crore in interest waiver losses and deferments alone. The total relief amount therefore would be around Rs. 20,000 crore (Rs. 13,000 crore from the CRF, NCCF and cost of foodgrain; Rs. 900 crore on payment of insurance claims; and 250 crore on emergency water supplies and the states own resources).

Drinking Water

During the 2002-03 drought, 18 states reported severe drinking water shortages; 11 of them continued to experience problems even the following year. In view of the serious drinking water situation, a separate task force for drinking water was set up under the chairmanship of the secretary, Ministry of Water Resources, in December 2002.

States were informed that they could avail of additional funds from the accelerated Rural Water Supply. Programmes (ARWSP). Five per cent of the funds under the program are earmarked for calamity relief and, accordingly, states were sanctioned Rs. 69.89 crore during the year. This is in addition to the regular flow of funds under the ARWSP.

Meanwhile, the Central Ground Water Board (CGWB) constituted its own task force and, during 2002-03, 66 of the board's drilling rigs were deployed across the drought affacted states. Over 250 wells were dug, of which 214 were functional. The board was instructed to hand the wells over to the state governments free of cost to ameliorate the drinking water shortage.

The states deployed water tankers in the rural hinterland. More than 40,000 with a capacity of 8,000 liters each, supposely made an average of four every day to the affected villages in the four peak summer months, carrying a total of 116 million kiloliters of water.

According to government estimates, the exercise benefited over 110 million including 22 million in urban areas. The entire tanker operation was filling through the NCF and the NCCF.

The railway network was also utilized tu supply water to Rajasthan an total of 2.9 million kiloliters of water were carried through 12 rakes (trains with 40 wagons of 20 kiloliter capacity) in the two states. The states provided free of cost.

Deployment of Tankers to Supply Drinking Water (Contingency Crop Plans)

State	*Number of tankers – Rural*
Andhra P-adesh	808
Chhattisgarh	NA
Gujarat	1,715
Karnataka	NA
Maharashtra	6,492
Orissa	6
Rajasthan	3044
Tamil Nadu	10640

Source: Department of Drinking water supply.

Livestock

Livestock plays a major role in the livelihood security of small and marginal farmers.

Any harm to live stock assets could have serious

implications on the ability of people to survive. Severe drought can decimate livestock populations.

Livestock numbers reduced due to drought or disease recover very slowly recovery of a herd could take several years, depending on the species. Herds of sheep and goat recover fast (around 25% growth a year), but cattle, buffaloes and camels grow slowly in numbers.

The Department of Animal Husbandry and Dairying (AH&D) at the Center, and its counterpart in the states, preside over issues relating to this sector. They are also responsible for the provision of adequate fodder, etc., to the farming community for the protection of livestock. The action plan for drought relief included measures to augment fodder in affected areas, distribute fodder at reasonable rates, monitor fodder prices and provide feed supplement for affected cattle.

Rajasthan and Maharashtra were the most severely hit with respect to number of livestock affected and the extent of fodder deficit. In Rajasthan, 226 lakh bovines were affected; Maharashtra followed with 189 lakh bovines. The fodder deficit in Maharashtra was worse than that of Rajasthan, at around 241.77 lakh MTs. For Rajasthan the figure was 149,31 Mts.

In view of the serious condition in Rajasthan, the task force on drought management released 30 MTs of damaged foodgrain for use as cattle feed. It was sold at Re 1 per kg to farmers after processing by the Rajasthan Cooperative Dairy Federation task force also decided to support cattle camps in the affected states. A 25 crore was made from the NCCF to assist NGOs in setting up cattle goshalas.

The Situation in Various States during 2002-03

State	*Number of bovines (in lakhs) affected*	*Fodder deficit (in lakh-MTS)*
Gujarat	54.84	2
Andhra Pradesh	153	61.61
Chhattishgarh	17.22	4.2

State	*Number of bovines (in lakhs) affected*	*Fodder deficit (in lakh-MTS)*
Karnataka	100	60.09
Madhya Pradesh	192	7.8
Maharashtra	189	241.77
Orissa	117	Surplus
Rajasthan	226	149.31
Tamil Nadu	86	Surplus

Source: Ministry of Agriculture.

At the height of the drought, 2,632 cattle camps sheltering and feed in animals were in operation across the states. The Indian railways pitch providing rakes to transport fodder.

Norms for relief expenditure under the head 'Cattle Care'

Item	*Scale*
For replacement of animals, mulch animals, or animals for haulage of livelihood	As per the pattern of subsidy under Swa Gram Swarojgar Yojana for small marginal farmers
For provision of fodder/ fodder concentrate	Large animals Rs. 12 per day Small animals Rs. 6 per day
Procurement, storage and movement of fodder	To be assessed by the National Center for Calamity Management (NCCM)
Movement of useful cattle to other areas	To be assessed by the NCCM for NCCF/ by state-level committee for CRF

Source: Department of agriculture and Co-operation, Ministry of agriculture

APPENDIX

A Chronology of Drought and Famine in India

Period	*Number of famines and Region droughts*	
5th century BC	1	Kashmir, Ayodhya (eastern UP), Rajasthan Maharashtra and Panjab
1sl century BC	1	Kashmir
9th century AD	1	Kashmir
10th century AD	2	Kashmir, Punjab, large parts of northern
11th century AD	2	Delhi, Bihar, Bengal, Orissa, Rajasthan,
13th century AD	4	Orissa, Bihar, Assam, Bengal, Rajasthan & Gujarat
14th century AD	5	Delhi, Rajasthan, Maharashtra, Mysore, Gujarat
15th century AD	5	Rajasthan, Maharashtra, Orissa, Assam parts of India, especially the areas along rivers Ganga and Yamuna
16th century AD	6	Most of the then Bombay presidency at Maharashtra, Delhi, Rajasthan, Kutch a central India.
17th century AD	6	Gujarat, Maharashtra, Rajasthan
18th century AD	24	Bombay presidency, Gujarat, Maharashtra, Madras
19th century AD	38	Rajasthan, Gujarat, Orissa, Maharashtra, Andhra Pradesh, Kashmir, Himachal Pradesh and Bombay
20th century AD	60	Rajasthan, Gujarat, Orissa, Andhra Pradesh, Himachal Pradesh, Bihar, Maharashtra, Kashmir

Source: 'Drgouht of Relief down to Earth', May 2001.

Administrative Districts Chronically Affected by Drought Conditions

State	*Districts*
Andhra Pradesh	Anantpur, Chittoor, Hyderabad, Karnool, Mehboob, Nalgonda, Prakasam
Bihar	Munger, Nawadah, Rohtas, Bhojpur, Aurangabad, Gaya
Gujarat	Ahmedabad, Amrely
Haryana	Bhiwani, Gurgaon, Mahendranagar, Rothak
Jammu and Kashmir	Doda, Udhanipur
Kanataka	Bangalore, Belgaum, Bellary, Bijpur, Chitraduraga, Chickmaga, Dharwad, Gulbarga, Hassan Kolar, Mandya, Mysore, Raichur
Madhya Pradesh	Betul, Datia, Dewas, Dhar, Jhabhua, Khandak, Shahdol, Shahja, Ujjain
Maharashtra	Ahmednagar, Aurangabad, Beed, Nanded, Nashik, Osmanabad, Parbhani, Sangli, Satara, Solapur
Orissa	Phulbani, Kalahandi, Bolangir, Kendrapada
Rajasthan	Ajmer, Banaswada, Barmer, Churu, Dungarpur, Jaisalrner, Jalo, Jhunjunu, Jodhpur, Nagaur, Pali, Udaipur
Tamil Nadu	Coimbatore, Dharmapuri, Madurai, Ramanathapuram, Salem, Tiruchirapali, Tirunelveli, Kanyakumari
Uttar Pradesh	Allahabad, Banda, Hamirpur, Jalan, Mirzapur, Varanasi
West Bengal	Bankura, Midnapore, Purulia
Jharkhand	Palamau
Chhattishgarh	Khargaon

Source "Drought: Assessment, Monitoring, Management and Resources Conservation." R. Nagarajan, Capital Publishing Co. Delhi

The most severe drought was witnessed in the state of Maharashtra in 1972 and 1987, 'moderate' to 'slight' drought is a recurrent feature. Parts of the state suffered drought in 2000, 2001 and 2003, continuing into 2004.

In October 2003 the government of Maharashtra declared that drought prevailed in 71 talukas of 11 districtsof the state: Pune, Satara, Sangli, and Sholapur (Pune division), Nasik and Ahmednagar Beed and Latur, Osmanabad and Aurangabad.

References

GOI, India Meteorological Department, Pune.

RBI, Annual Report 2001-02 Mumbai.

Relief Administration in India.–Country Report 1999, National Disaster Management Division, Ministry of Agriculture.

I Putnaik, Drought Proofing India, The Indian Express, August 9,2004.

GOI, Economic Survey 2002-03, Ministry of Finance.

Drought of Relief, Down to Earth, May, 2001.

B.M. Bhatiya, Famines in India, 1770-1943, Asia Publishing House Mumbai, 1958.

Mohan Rakesh, A Decade after 1991, New Challenges Facing the Indian Economy RBI, Bulletin, November 2002.

RBI, Report on Currency and Finance 2001-02, Mumbai.

RBI, Report on Currency and Finance 2002-03, Mumbai.

RBI, Annual Report 2002-03, Mumbai.

Administrative Staff College of India. A Study of the Fiscal and monetary Implications of Excess Stocks of Foodgrains, Hydrabad, 2002.

GOI, Economic Survey 2002-03, New Delhi 2003.

Hindu Business Line, Service Sector Growth, What does it Mean? 2002.

N. Nagarjan, Drought; Assessment, Monitoring, Management and Resource Conservation, Capital Publishing Co., Delhi.

7

Conclusion

It is true that the famines in India often begin with the failure of monsoons. But it cannot be the only one cause of famines, or deficiency of rainfall in one year did not inevitably lead to famine, which implies 'want of food, hunger and starvation'.

Indian political leaders like Dadabhai Naoroji and economist R. C. Datta had blamed the British administration and particularly its system of taxation or revenue policy for the fast developing poverty of the people. There had been also a severe decline of traditional Indian industries. In consequence there was widespread unemployment. Landless households had no other source of income as the cottage industries disappeared; they were left destitute. Dr. Srivastava has rightly come to the conclusion that the increasing poverty of the people was a very important cause of the growing intensity and widening incidence of Indian famines.

The Famine Commission of 1880 saw the loss of employment during droughts as the main problem and noted that food was rarely in short supply for the country as a whole. The self-selection feature of relief works was relied upon to ensure financial economy while providing a guarantee against starvation. It would be simplistic however to regard to days relief system as a mere legacy of the British administration. In fact, important changes have occurred since independence which marks the second milestone of the recent history of famine relief in India. In particular, the resources allocated to relief have increased, and the range of measures they support has broadened. The government has entered the previously sacrosanct domain of food supply management. Last but not least, the weak commitment of the British administration to the prevention of famines had evolved into political compulsion to respond to public pressure. But the disappearance of large scale famines in India is coexisting with the persistence of mass poverty and hunger.

Today 40 per cent population (especially rural) is not able to obtain a subsistence income even in normal times, because of unemployment, under employment and low wages. This results in inadequate food, malnutrition, diseases.

Government has been spending crores of rupees for relief measures. For example, the Government of India spent Rs. 315 crores on D.P.A.P., during the Sixth Five Year Plan. But all relief work is on adhoc basis or in the form of adhoc schemes like Employment Guarantee Scheme, Small and Marginal Farmer Agency and Drought Area Development Programme, N.R.E.P., R.L.E.G.P. etc. In fact most of these schemes do not reach the very poor, they remain unemployed, and therefore chronic poverty persists.

The other problem is that of production and supply of foodgrains. India faced shortage of foodgrains and had to import grain till the seventies. The level of production increased substantially in the post green revolution period, but instability in agricultural production has increased on account of high sensitivity of the new varieties to variations in rainfall. This rising vulnerability of agricultural output, especially foodgrains output, to droughts is traceable to the high complementarity of new seed fertilizer technology with water and the inadequate expansion of irrigation facilities. Although area under irrigation has increased from about 17 per cent of the cultivated area in early 1950s to around 30 per cent by 1987, a good part of this irrigation is itself dependent on rainfall.

The severity of successive droughts has increased on

account of deforestation. While there is no clear evidence of change in rainfall pattern on account of deforestation, the impact of the latter in terms of depletion of soil moisture, lowering of water table because of run-off of rain water and shortage of fuel wood and fodder is well established.

The causes of deforestation are rooted in the pattern of development and management of resources which has led to commercial felling of trees on a massive scale to meet the industrial and commercial demand.

Ground-water has also been indiscriminately exploited in India for irrigation and therefore ground-water level is going down in many parts, and thus the problem of drought, particularly that of shortage of drinking water is becoming critical. In addition problems of water logging and salivation are being faced in the irrigated tract and consequently growth in productivity levels cannot be maintained.

The green revolution technology did not take into consideration the ecological consequences of mono-hybrid culture, and intensive use of fertilizers, pesticides and irrigation. The country is now facing the ill consequences as soils are turning unproductive, pests are turning resistant to pesticides, water bodies are getting polluted and fish are dying. As indicated in Chapter 4, it is now being realised that we should not put emphasis on chemical inputs but on eco-biological balance; and that the problems of attaining higher and sustainable yields and of providing more employment achieving higher productivity progressively have to be handled simultaneously so that the chronic famine of work and famine of food, fodder, fuel and drinking water are eliminated.

Bibliography

Aykroyd (1974). *The Conquest of Famine,* Chatto and Windys, London.

Baird Smith (1861). Report on the North Western Provinces Famine of 1860-61.

Baishya, P. (1975) *'Man Made Famine,'* Economic and Political Weekly 10(21), pp. 321-822.

Bandhyopadhyay, J. (1987). *Political Ecology of Drought and Water Scarcity: Need for an Ecological Water Resources Policy,* E.P.W. 22:50, 12th December, 1987.

Bandyopadhyay, P. (1987). *Indian Famine and Agrarian Problems,* Star Publication, Calcutta.

Bardhan, K. (1977). *Rural Employment, Wage and Labour Markets in India.*—A Survey of Research, E.P.W. 12.

Bhatia, B. (1962). *Famines in India – A study in Some Aspect of the Economic History of India (1860-1945),* Asia Publishing'House, Bombay.

Brahme Sulabha (1983). *Drought in Maharashtra 1972—A Case for Irrigation Planning,* Gokhale Institute of Politics & Economics, Pune.

Brahme, Sulabha; Pore, Kumud (1990). *Land Use in Western Zone of India : Problems and Prospects,* Gokhale Institute of Politics & Economics, Pune.

Borkar, V. V. and Nadkarni, M. V. (1975), *Impact of Drought on Rural Life,* Popular Prakashan, Bombay.

Central Institute of Research and Training in Public Cooperation (1969). Famine Relief in Bihar : A Study, C.I.R.T.P.C., New Delhi.

Chetty, V. and Ratha, D. (1987). *The Unprecedented Drought and Government Policies,* Indian Statistical Institute, New Delhi.

Dandekar & Rath (1971). *Poverty in India,* E.P.W., Vol.VI, No. 2, 9 June.

Datta, Romesh (1878). Indian Famines their Causes and Prevention Underdeveloped Wealth in India and State Reproductive Works, the Ways to prevent famines and Advance the Material Progress of India, London.

Deccan Riots Commission Report 1878.

Dhawan, B. (1987), The Drought of 1987, Impact on Foodgrain Output, *Economic Times,* 17th August.

Digby, W. (1878). *Famine Campaign in Southern India,* Longmans, Green and Company, London.

Dreze, J. (1988). *Famine Prevention in India,* The Development of Economic Research Programme, London.

Dutta, R. C., *The Economic History of India in the Victorian Age,* Fourth ed., London.

Dutta, R. C., (1900). *Famines and Land Assessment in India,* London.

Dwivedi, D. (1974). *Economic Concentration and Poverty in India,* Datta Book Centre, Delhi.

Etheridge, A. (1868). *Report on the Past Famines in the Bombay Presidency,* Education Society's Press, Bombay.

Gadgil, D. P. (1929). *Industrial Evolution of India,* Oxford University Press.

Gangrade, K. D. and Dhadda, S. (1973). *Challenge and Response: A Study of Famines in India,* Rachana Publications, Delhi.

Government of Bihar (1973). *Bihar Famine Report 1966-67,* Secretariat Press, Patna.

Government of Maharashtra (1973). *Report of the Fact Finding Committee for Survey of Scarcity Areas of Maharashtra* (Sukhantankar Committee), Bombay.

Government of India (1886). *Report of the Indian Famine Commission 1880,* London.

Government of India (1898). *Report of the I.F.C. 1898,* Simla.

Government of India (1901). *Report of the I.F.C. 1901,* Calcutta.

Government of India (1931). *Central Banking Enquiry Committee.*

Government of India (1945). *Famine Inquiry Commission Report on Bengal,* New Delhi.

Government of India (1972). *Report of the Task-Force on Ground Water Resources.*

Government of India (1985). *Annual Report of the Ministry of Rural Development.*

Government of India. *Statistical outline 1986-87.*

Government of India. *Sixth Five Year Plan.*

Government of India. *Seventh Five Year Plan.*

Government of India (1986), *Economic Survey 1986-87.*

Government of India (1987). *Economic Survey 1987-88.*

Government of India (1987). *Implications of 20 Point programme,* Quarterly Progress Report, June.

Greenough, Paul R. (1982). *Prosperity and Misery in Modern Bengal: The Famine of 1943-44,* Oxford University Press, Oxford, New York.

Hanumantrao, C. H.; Ray, S. K, and Subbarao, K. (1988). Unstable Agriculture and Droughts, Vikas Publishing House.

Hegde, R. (1987). Drought conditions in India Needs a New Approach in Financing the Relief Measures, Southern Economist, 26, 13 November.

Jathar and Beri (1939). *Indian Economics,* Oxford University Press, 1939.

Jodha, N. S. (1975). Famine and Famine Policies Some Empirical Evidence, E.P.W. X(41): 1609-1623.

Journal of Facts for You, *Monthly Economic Affairs,* June 1988.

Kaushal, G. (1979). Economic History of India, 1757-1966, Kalyani Publishers, New Delhi.

Krishnamachari, K.A.V. et al. (1974). Food and Nutritional Situation in the Drought affected Areas of Maharashtra, A survey, The Indian Journal of Nutrition and Dietetics.

Liberman, S. (1984). Field Level Perspectives on the Administration of Maharashtra' E.G.S., Working Paper No. 109, The Population Council, New York,

Loveday, A. (1914). *The History and Economics of Indian Famines,* A. G. Bell and Sons, London.

Maloo, K. (1987). The History of Famines in Rajputana (1858-1900 A.D.), Himanshu Publication, Udaipur.

Masefield (1963). *Famine: Its Prevention and Relief,* Oxford University Press, Oxford.

McAlpin, M. (1983). *Subject to Famine : Food Arises and Economic Change, in Western India 1860-1920,* Princeton University Press.

Munel (1985). *Drought Prone Area in India,* B.S. 14 February.

Naoroji, D. B. (1901). Poverty and Un-British Rule in India, Swan Sonnenschein and Company, London.

Pandey, S. M. (1979), *Effects of Droughts on Rural Population, Findings on Area Study,* Indian Journal of Industrial Relations, 15(2), pp. 251-305.

Rajadhyaksha, Radha. (1987). *Hunger,* The Journal of Third World, Nov/Dec., No. 11.

Ray, P. C. (1901). *Indian Famines, Their Causes and Remedies,* Cherry Press, Calcutta.

Ray, S. C. (1909). *Economic Causes of Famines in India,* Baptist Mission Press, Calcutta.

Saibaba, G. (1986). Planning and Management of Droughts, Journal of Khadi Gramodyog, February.

Singh, J. (1985). "Some Aspects of the Strategy of Agricultural Development in Eastern Region," Unpublished paper reported in Appendix III of the Report of Study Group on agricultural strategies for Eastern region of India, G.O.I., Planning Commission, New Delhi.

Singh, K. Suresh (1975). The Indian Famine, 1967: A Study in Crisis and Change, Peoples Publishing House, New Delhi.

Singh, Tapeshwar (1978). Drought Prone area Programme in India, Aspect of Identification and Development Strategy, Peoples Publishing House, New Delhi.

Srivastava, H. (1968). *History of Indian Famines and Development of Famine Policy* 1858-1918, Sri Ram Mehra and Company, Agra.

Subramanian, V. (1975). *Parched Earth, the Maharashtra Drought 1970-73,* Orient Longmans, Bombay.

Vakil, C. N. (1924). *Financial Development in India 1860-1924,* Bombay.

Velyudhana, T. (1987), *Monsoon and Economy,* R.B.I. Annual Report 1986-87, Sept., E.P.W., Vol. XXIII, No. 49, December.

Venkatachary, R. (1987). Combating Drought: Long and short term plans, *Economic Times,* September.

Vijayaraghwan, R. (1987). *Drought Relief Financing, Business Standard,* September.

Walkar, P. S. and Singh, R. A. (1986). *Risk Benefits, Crop Insurance and Dry Land Agriculture,* E.P.W., June.

Yojana (1985). *Drought and Management* by Basu Srelekha.

Index